Transcending the Economy

Transcending the Economy

On the Potential of Passionate Labor
and the Wastes of the Market

Michael Perelman

St. Martin's Press
New York

TRANSCENDING THE ECONOMY

Copyright © 2000 Michael Perelman. All rights reserved. Printed in the United States of America. No part of this book may be used or reproduced in any manner whatsoever without written permission except in the case of brief quotations embodied in critical articles or reviews. For information, address St. Martin's Press, Scholarly and Reference Division, 175 Fifth Avenue, New York, N.Y. 10010.

ISBN 0-312-22977-1

Library of Congress Cataloging-in-Publication Data
Perelman, Michael.
  Transcending the economy : on the potential of passionate labor and the wastes of the market / by Michael Perelman.
     p.   cm.
  Includes bibliographical references and index.
  ISBN 0-312-22977-1
  1. Economics.   2. Waste (Economics).   3. Human capital.   4. Social justice.   5. Waste (Economics)—United States.   II. Title.
HB71.P47   2000
335.5—dc21   99–055566
                                                                              CIP

Design by Letra Libre, Inc.

First edition: April, 2000
10 9 8 7 6 5 4 3 2 1

Contents

Preface

"Where there is no vision, the people perish." Proverbs 29: 18

This book is an exploration of potential. You may wonder why an economist would be writing about potential. In a sense, an exploration of potential is the opposite of economics, which is, for the most part, the study of limits. Economics purports to give a scientific explanation for all the reasons why we cannot dare to go beyond the status quo of the market system.

Economic theory teaches that societies that fail to follow the rules of the market will fall short of this potential, but the market sets the outermost limit of what society can obtain. Through intensive application of education and technology a society can modestly expand that limit by a percent or two per year, but only if it follows the dictates of the market.

I take issue with the vast majority of my fellow economists. I am convinced that the economy, at least in its present form, comes nowhere near its full potential. More important, a market society cannot not provide satisfactory lives for most of its participants. Still more important, markets lead to environmental disruption, social discord, and wasted opportunity.

Many people have an interest in expanding their own personal potential. Bookstores are filled with shelves of publications promising methods of increasing personal potential. The general public has a fascination with underdogs, overachievers, and heroes who manage unexpected feats. The media play upon this desire for transcendence by selling images of extraordinary people who perform the impossible. Unfortunately, our cinematic heroes do little to inspire us with a sense of our own latent potential. Instead, they mostly confirm that these exceptional achievements are beyond the reach of ordinary mortals.

Yet, when we look around us we can find innumerable examples of people who actually have transcended their circumstances by successfully battling against physical limitations, social conditions, and even historical circumstances. What one person can do can be impressive, but such achievements do not come anywhere near what is possible when a whole society sets out to achieve a shared goal.

Cynics contend that we cannot risk embarking on untried experiments. I take some comfort in the words of Alfred Marshall, surely the most influential economist of the early twentieth century, who wrote, "No doubt men, even now, are capable of much more unselfish service than they generally render: and the supreme aim of the economist is to discover how this latent social asset can be developed most quickly, and turned to account most wisely" (Marshall 1927, p. 9). Although his own pedestrian economics failed to match his soaring rhetoric, what Marshall advocated three quarters of a century ago is more true than ever today.

This book will refute the objections of those who warn that the risks of attempting to transcend the present economic arrangement are too great. In fact, our society presently has enormous leeway in the form of wastes—foremost of which, as Marshall recognized, is the waste of human potential. Probably less than one-quarter of all economic activity actually contributes to human welfare.

With all this slack in the economy, any new form of social organization, especially if society can establish a store of goodwill along the way, means that we have the potential to exceed by leaps and bounds anything the world has ever seen. Moreover, the real risk is complacency. If we allow society to continue on its present course we risk destroying the environment upon which we all depend.

PART 1

Waste of Nations

CHAPTER 1

Introductory Section

Deprivation Chamber

A sensory deprivation chamber is a dark, soundproof room in which a solitary person cannot hear anything or see anything. After a short period in one of these rooms, a person begins to feel soreness.

From where do these aches come? Normally, physical imbalances in the body cause some muscles to fight against other muscles. We are not usually aware of this fatigue. Although we may not be conscious of these tensions, they may eventually lead to acute physical problems.

Outside of the sensory deprivation chamber, this tension remains in the background. We are usually preoccupied with the onrush of ordinary events. Inside of the deprivation chamber, we no longer have the typical stimuli that distract us from becoming acutely aware of our physical state.

So, cut off from these outside concerns, we take closer note of the state of our body. This experience will not cure the ills that afflict a person, but it can provide some clues about what could possibly lead to improvement.

The sensory deprivation chamber helps us to understand something about a recurrent mystery of human activity. Every so often we hear of somebody performing an extraordinary feat. For example, I keep a clipping on my desk that describes a young mother whose car accidentally rolled onto her child. Somehow, the mother was able to lift the vehicle off the child.

When I think of this event, I imagine that this same woman may have just recently asked her mate to help her carry some heavy groceries from the car. Suddenly, when faced with an immediate emergency, she can lift up an automobile.

How is this mother's action possible? Had she given the situation a moment's thought, she would have considered the task impossible, yet somehow without a moment's thought she lifted the vehicle.

Can we just ascribe the mother's effort to a rush of adrenaline? I believe that something else is at work. Remember how our muscles customarily work against each other. In the midst of an emergency, all this wasted effort comes to a temporary end. All of a sudden, the mind is focused on a single objective. In the case of the young mother, all of her muscles kicked in and began to work in tandem, and then, with seemingly superhuman force, she lifted the car.

I am writing this book in the hopes you can experience it something like a social sensory deprivation chamber. If I am successful in this endeavor, you will get a sense of how our society dissipates the efforts of those who work, while remaining unaware of the enormous potential that lies within each and every one of us. Even more important, I will show how an intelligent organization of society can magnify the collective potential far beyond the sum of individual potentials. In the process, I hope to provide a panoramic view of the innumerable forms of waste that a rational society could avoid.

In addition, I would like to convey a feeling for the enormous harm that our present day society inflicts on people. I will direct your attention to the manner in which the defects in our economy contribute to the social pain that people inflict on each other—whether they be rich or poor, black or white. You will get a sense of the exhaustion that this social tension creates for all of us, although in a society such as our own, this tension will probably be as unequally distributed as wealth.

Finally, I would like to suggest that, just as the mother was able to lift the car off of her child, our society could lift the weight from people so that they could achieve far more of their potential than anyone imagines.

I do not intend to recommend specific cures for social ills; rather I think of my project as comparable to the work of Sadi Carnot (1796–1832), the early nineteenth-century French scientist who analyzed the nature of an ideal engine. He realized that such an ideal engine, now known as a Carnot engine, would be every bit as impossible as the creation of an ideal society. Nonetheless, an understanding of an ideal machine does have a value. By comparing the performance of a real machine with an ideal machine, we get a measure of efficiency.

Unlike the physicists who can make precise calculations for an ideal machine, I have no comparable formulae to allow me to make an exact analysis. Instead, I can only discuss those aspects of our present economy that seem to indicate waste and inefficiency.

The wastes of our own society far exceed the inefficiencies of an ordinary engine. Unlike the Carnot engine, which must work within the strict limits of physical laws, people, such as the woman who lifted the car, have incredible untapped potential that can permit a community to develop unimaginable possibilities. Perhaps the person that came closest to recognizing this potential was Charles Fourier (1772–1837), a brilliant, but eccentric French writer who introduced the concept of passionate labor, in speculating about the ideal society around the same time that Carnot was working on his ideal engine.

I can only hope that, after reading this book, you will have an intuition of the capacities of a coherent society—one in which the various parts reinforce each other rather than work against each other. At that point, we can get together to begin to work toward the creation of a world that, while it will not be perfect, at least it will be a far better society than the one that we live in today.

Beyond Economics

Economies, like people, rarely work to their full potential. Even so, our society indoctrinates us to think of markets as the natural state of affairs. According to this perspective, any attempt to transcend the market is doomed to failure. This book is about the wastes associated with markets and even more about the transformation of the economy into a society in which people have the opportunity to achieve their full potential.

The first part of this book will discuss waste. I do not pretend to provide a full compendium of wastes, but merely a survey of a set of representative wastes. Although I will make the case that the elimination of waste has the potential to provide vast improvements in people's wellbeing, an emphasis on the elimination of waste promises both too much and too little. Of course, we could never expect to eliminate all of the wastes in any economy, let alone one structured as our own is. In this sense, an emphasis on waste promises too much.

In another, more profound sense, an emphasis on waste promises far too little. Even if we could somehow eliminate all the waste in our society, we still would not bring our economy to anywhere near its full potential.

At this point, we must drop our earlier metaphor of the Carnot engine. We cannot merely postulate the principle of an ideal society and then go about constructing it. Of course, correcting even a few of the wastes and inefficiencies that I will discuss later could provide significant amelioration.

To achieve true efficiency, not just mere economic efficiency, requires a fundamental transformation of society. At the heart of this transformation is an entirely different conception of human nature. Of course, this true efficiency transcends the usual utilitarian notion of efficiency. Rather, this would permit members of society to be able to enjoy the fullness of their human potential. Our usual measures of efficiency, such as profit and loss or dollars and sense, would have little, if any, relevance in such a world.

Before we make that leap in thinking about efficiency, let us adhere to the more conventional conception of efficiency. This book will begin by suggesting the possible improvements in economic performance that the elimination of wastes would permit.

Economic theory rules out the possibility of such a wholesale remake of our economic organization. Instead, economics insists on a stunted vision of human potential, based on the unworldly assumption that people behave in accordance with the confining concept of economic man. Sadly, a great deal of behavior in our society does seem to confirm the deeply-held assumptions of economic theory.

Unfortunately, this economic man who stands at the core of economic theory is a pitiful, deformed creature. Lacking outlets for true creativity and accomplishment, he turns to the mindless accumulation of wealth and power. In contrast to the sensory deprivation chamber that makes people become acutely aware of underlying discomforts, the endless supply of goods from shopping malls serves to numb the realization of emptiness.

If economic man cannot make anything of his life, he will make damn sure that others will find themselves in even less desirable circumstances. However much satisfaction the dire straits of others might give him, somebody somewhere will occupy a superior position. This knowledge will gnaw at him, driving him further and further into the mad scramble to accumulate wealth. This pathological behavior leads to social and environmental mayhem.

Rather than accepting the possibility that alternatives to this restrictive view of the world might exist, either in theory or in reality, economic theory confidently presents itself as a bearer of absolute truth. As Kenneth Arrow, himself a Nobel laureate in economics, once wrote, "An economist by training thinks of himself as the guardian of rationality, the ascriber of rationality to others, and the prescriber of rationality to the social world" (Arrow 1974, p. 16).

This economic rationality is merely the rationality of markets. Since markets have played such an inordinate role in our society, to go beyond economics means that we must venture into uncharted territory. Yet to fail to do so would condemn the majority to a life of misery and despair.

The idea of transcending the market might seem far-fetched to economists. Their models assume some sort of efficiency or optimality. Their understanding of human psychology implies that markets remain the very best that any society can accomplish, although more "liberal" economists allow that we can modestly tinker with markets to achieve a bit of an improvement in economic performance.

The United States Economy as an Ideal System

Let me propose a simple thought-experiment. Returning to my comparison with Sadi Carnot, suppose we were to look at human society from the perspective of an engineer who is analyzing the efficiency of a machine. Yes, I know that my thought-experiment smacks of a cold indifference to everything human. I fully realize we are not machines, but bear with me for a moment.

In a sense, my experiment is not novel. Economists often try to explain why other economies fall short of the performance of the United States. For example, Charles Jones and Robert Hall (1999) note that in 1988, output per worker in the United States was 35 times greater than output per worker in Niger.

Jones and Hall estimate that the difference in capital intensities made workers in the United States one-and-one-half times more productive than workers in Niger. According to their estimate, educational differences could make workers in the United States 3.1 times more productive again. Still, accounting for the combined effect of these two traditional explanations of productivity differences would leave workers in the United States only 7.7 times more productive than those in Niger, far less than the actual multiple of 35.

Jones and Hall attribute the shortfall in Niger's performance to what they call the social infrastructure—the institutions and government policies that determine the economic environment within which individuals accumulate skills, accumulate capital, and produce output.

Obviously, the conventional measure of output per worker is higher in the United States than in Niger. In this sense, the economic performance of the United States provides a convenient bench mark. Even so, I see no reason why we should not subject the economy of the United States to the same degree of scrutiny that we apply to less developed economies, such as Niger. After all, what reason do we have to believe that the United States represents the outer limit of economic and social possibilities?

Of course, I would not want anyone to take my engineering metaphor too literally. The attempts of strong, centralized states to engineer their societies often fall painfully short of the original vision, just because they seek to manipulate their society from above rather than engaging the populace. Later, in Chapter 9, I will return to the pitfalls in attempting to engineer society.

Societies, like natural environments, are complex arrangements with far too much subtlety to be engineered from above. Recently, James Scott, Eugene Meyer Professor of Political Science and Anthropology at Yale University, has published a searing theoretical critique of social engineering (Scott 1998). While some of his specific examples are not entirely convincing, his abstract analysis is rock solid.

As we will see in Chapter 10, the key to success in transforming society lies in engaging people with their informed consent. By engaging, I do not merely mean browbeating or propagandizing people. Their consent must be informed. More important, the engagement cannot be a top-down system of command and control. Engagement means that the social vision must percolate up more than it permeates below.

An Initial Indication of Waste in the United States

What sort of factors might be holding back the potential of the United States economy? Some of the causes of the shortfall in the economic performance of the United States are fairly self-evident. Just think how much time and energy is wasted in attempting to manipulate people who have good reason to resist? Advertisements endeavor to induce us to feel a desire for a product that we might buy cheaper elsewhere or do not even need at all. How much effort is lost by workers who intentionally fail to do their best or even resist outright just because they resent their treatment? Other people know that they could lose their job if they perform it too well by making themselves obsolete.

In addition, most people seem to be discontented with their roles as workers, consumers, or citizens. Our imaginary engineer would have to conclude that this machine does not work up to its potential. It has too many leaks, too much friction, and too little satisfactory output considering the enormous quantity of resources consumed.

The environmental shortcomings of the United States are no less obvious. For example, the United States economy spewed out 209 million tons of garbage in 1995, or a bit less than one ton per person according to data from the Environmental Protection Agency (1999). Doug Henwood, editor of the invaluable *Left Business Observer*, informed me that New York City alone produces 26,000 tons of garbage every day (New York City Department of Sanitation). These figures exclude the innumerable tons of garbage strewn about as litter, which never arrive at an official garbage dump. Even more ominously, the economy bombards the environment with enormous quantities of toxic waste products. The more affluent communities protect themselves to some extent by shipping their hazardous waste products to places where their poorer brethren reside—perhaps even in Niger. In fact, wunderkind economist, nephew of two Nobel prize winners in economics, and now Secretary of the Treasury, Lawrence Summers, wrote an infamous memo when he was vice president of the World Bank, suggesting that such "countries in Africa are vastly under-polluted" in the sense that affluent nations, such as his own, can afford to pay the poor in such societies to accept their toxic waste products (Anon 1992).

The excessive demands on our time represent another dimension of waste. Some years ago, Juliet Schor created a stir by showing that people in the modern United States were working more and more hours (Schor 1991). In addition, the increasing time consumed by commuting eats up even more of the working day. For example, people in the United States spend 8 billion hours a year stuck in traffic. In cities with populations of more than 1,000,000, the average commute time is 31.9 minutes each way (Glaeser 1998, p. 151). Each year, traffic jams get worse. The costs of this time range from $43 billion—according to the Federal Highway Administration—to $168 billion in lost productivity estimated by other economists (Kay 1997, p. 121). These estimates do not include the toll that this congestion takes on people's mental and physical health.

For many people, the working day is becoming more intense. Faxes, e-mail, and cell phones continually demand our attention. At the same time, less and less work is required to produce the goods and services that we enjoy.

What is happening? An increasing amount of time and effort is devoted to activities that add little or nothing to the total product, so much so that our work demands increase despite the shrinkage of the necessary work demands.

The greatest waste concerns the fate of those whom society treats as human detritus: the unemployed and the supposedly unemployable. Contemporary society treats the fate of such people as evidence of their own personal dereliction. As the philosopher, Hannah Arendt, once observed: "those who are completely unlucky and unsuccessful are automatically barred from . . . the life of society. Good fortune is identified with honor, and bad luck with shame. . . . The difference between pauper and criminal disappears, both stand outside society" (Arendt 1966, pp. 141–42). In contrast, I regard the condition of these people as a striking proof of the failure of society. One of the few economists who shared this perspective was Alfred Marshall, who before he became the preeminent economist of the Anglo-Saxon world, declared:

> in the world's history there has been one waste product, so much more im-
> portant than all others, that it has a right to be called THE Waste Product.
> It is the higher abilities of many of the working classes; the latent, the un-
> developed, the choked-up and wasted faculties for higher work, that for lack
> of opportunity have come to nothing. Many a fortune has been made by
> utilizing the waste products of gas works and of soda works; it has been very
> good business. But a much greater waste product than these is the founda-
> tion of the fortunes of co-operation. Let us take stock of the resources of
> co-operation in this country. (Marshall 1889, p. 229)

Waste Versus the Surplus

For hundreds of years, observant people have noticed that a relatively small number of workers sufficed to produce the goods and services that support the rest of society. For example, the ever-practical Benjamin Franklin noted:

> It has been computed by some political arithmetician . . . that if every man
> and woman were to work for four hours each day on some thing useful, that
> labour would be sufficient to procure all the necessaries and comforts of life,
> want and misery would be banished out of the world, and the rest of the

twenty-four hours might be leisure and pleasure. (Franklin to Vaughan, 26 July 1784; in Franklin 1905–07; 9, p. 246)

The flip side of Franklin's observation was the profusion of professions and occupations that seemed to make no economic contribution to society. For example, Franklin's contemporary, Sir William Petty, often considered to be the first major political economist, lumped together as unproductive: "thieves, robbers, beggars, fustian and unworthy Preachers in Divinity in the country schools. . . . Pettifoggers in the Law. . . . Quacksalvers in Physick, and . . . Grammaticasters in the country schools" (cited in Strauss 1954, p. 137).

During the early nineteenth century, a group commonly mislabeled as Ricardian socialists—after the British economist, David Ricardo (1772–1823)—though they were neither Ricardian nor socialist, carried on with the tradition of making empirical estimates of the ratio of necessary labor relative to the labor of those whose made no material contribution to the goods and services on which we depend (see King 1983). One who wrote under the pseudonym, Piercy Ravenstone, declared: "Anciently one man's labour was sufficient to maintain two families; in France it is sufficient to maintain more than three. In England it is equivalent to the subsistence of five" (Ravenstone 1821, p. 221; cited in King 1983, p. 350). The precision of Ravenstone's numbers may be suspect, but he was certainly correct in one respect: even if more than a quarter of the work day was necessary in Franklin's day, a small fraction of that quarter would suffice today. Thus, if everybody did productive work, we could cut the workday by far more than three-quarters.

When earlier commentators discussed the superfluous people, they usually based their calculations on the concept of productive labor. This category was very elastic, depending upon the person who was making the calculation. Productive labor could include just about everybody, except lawyers and bankers. Later, as the definition of unproductive labor became more standardized, workers engaged in commercial or trading activities fell into the category of unproductive labor (see Moseley 1991).

Fred Moseley, an economist at Mount Holyoke College, has done more work analyzing the dimensions of unproductive labor than anybody else. His numbers are less striking than those of Franklin because he is only looking at a subset of the total labor force. Even so, according to Moseley's calculations, the ratio of unproductive labor to productive labor increased by 83 percent, from 0.35 in 1947 to 0.64 in 1977 (see Moseley 1991, table 4.3 and pp. 111–15). No doubt, this proportion has increased significantly since 1977.

Anwar Shaikh of the New School and Ahmet Tonak of Bard College made some calculations that throw additional light on the extent of unproductive labor. They estimated from the statistics reproduced in the National Accounts that 41,148,000 American workers produced goods valued at $4,149.75 billion in the year 1989. Of this product, they consumed goods to the value of $1,206.4 billion, leaving a surplus product of $2,943.35 billion. In other words, each of these American workers produced goods to the value of around $100,000, and took home just under $30,000 while contributing $70,000 to the surplus (Shaikh and Tonak 1994, Table 5.6, p. 117).

My concept of waste goes considerably further than these calculations of unproductive labor. I will discuss how considerable wasted effort occurs, even within the sphere of productive labor. For example, productive labor can produce goods that serve no social need, such as speculation. As a result, some of the labor that goes into the production of commodities, such as paper, computers, or buildings, turns out to be unproductive. Besides, even within the sphere of productive labor, the economy dissipates its energies in unnecessary work, as well as in losses due to conflict and bad morale in the workplace.

I adopt an approach similar to the way that scientists help a runner to achieve maximum speed. These scientists create a computer display to show the runner all the ways in which she or he wastes effort in the act of running. Is there a slight lateral movement in the knee? Is the foot hitting in just the right place? Of course, the scientists already have a fairly precise model of the ideal body mechanics.

My task is far more complex. First, I fully realize that no society could ever eliminate all of the wastes that I will discuss. Second, any interpretation of waste will have an unavoidable element of subjectivity. Finally, since a model of an ideal society does not yet exist, I will be unable to quantify the exact dimension of the waste involved. As a result, my efforts will have some obvious limits.

Sometimes, a discussion of waste is intended as a prologue to a call for more regimentation in the name of efficiency. I draw a very different lesson from this exercise. I will indicate why regimentation will lead to a less efficient (as well as a less pleasant) society, simply because more regimentation, at least when it is imposed from above, breeds more conflict.

Rather than calling for more regimentation, I believe that the staggering quantities of wasted effort in our society that I document below offer a wonderful opportunity. Since so little effort is required to do what society already does, we can safely get on with the business of restructuring society. In the process, we can redirect the effort that we now dissipate in waste to create forms of production that allow us to tap into our inherent creativity.

A Brief Mention of Passionate Labor

Any discussion of waste is necessarily limited. We can imagine our existing economy working under maximum efficiency as the economic equivalent of a Carnot engine. Any dropoff from that level of efficiency could indicate the existence of waste.

Unlike the Carnot engine, which has a maximum potential efficiency, as I mentioned earlier, human activity is indeterminate. While the laws of physics might tell us about the maximum possible weight that the human body could lift, we have no standard for measuring the latent potential for ingenuity and creativity. People can call upon initiative, creativity, and sheer determination to accomplish feats that might otherwise seem unimaginable. Nobody would have thought that the woman in the press clipping had the capacity to have lifted the car.

During the middle of the nineteenth century the utopian, Charles Fourier, wrote about what he called "passionate labor." Unlike individuals faced with the necessity of performing drudge work, people engaged in passionate labor approach their tasks with enthusiasm and joy.

To create a society in which passionate labor becomes the norm requires a wholesale transformation in the way that we organize our lives. We would have to go about demolishing systems of domination, class, and the great divisions of rich and poor. I do not pretend to know how to accomplish such lofty goals; I will only direct your attention to some indications of the possibilities associated with passionate labor.

If we contrast the current economy with what we could accomplish under a system in which passionate labor became the norm, present economic achievements would appear trifling. We will get hints of this potential later when we look at economic performance during times when society becomes less divisive, such as during times of war or other national emergencies.

Our discussion of the possibility of passionate labor will necessarily be relatively brief since we must rely mostly on anecdotes and speculation. Nonetheless, the potential of a regime of passionate labor is so enormous that this subject demands our attention.

For now, we will drop the subject of passionate labor and return to the subject of economic waste.

CHAPTER 2

Taxes, War, and the Elimination of Waste

Flat Tax and the Reduction of Waste

Since I am going to begin analyzing how the economy dissipates time and resources, let me begin with a brief discussion of a subject upon which I would expect to find general agreement: taxes. Nobody likes them. Forget about the inequities. For now just think about the enormous waste of time and resources associated with keeping records, filling out forms, staffing of the huge bureaucracy charged with enforcing the collection process, as well as the socially unproductive nature of the widespread efforts to avoid taxes.

Economists refer to the costs of all the activities listed above as dead-weight losses, because they add to the burden of running an economy without any corresponding social gain. In other words, society does not have any more food, clothing, or shelter available because of the activities of people engaged in such activities.

Tax preparations alone represent a substantial economic sector. Lawyers and accountants also devote uncountable hours to devouring thick volumes of tax code so that they can devise brilliant methods of circumventing the law. The manipulation of economic activity to avoid taxes imposes an even greater burden on the economy. Farmers decide what crops to plant, corporations relocate factories, and all the while money zips around the globe, at least in part, just to avoid taxes.

We can get an image of the enormity of the business of tax avoidance by considering the rise of international tax havens. Perhaps the most notorious tax haven is the Grand Cayman Islands, with a population of a mere 30,000. This tiny land happens to be the fifth-largest nation in the world in terms of booking bank loans (Greider 1997, p. 32).

The islands have almost as many registered companies (25,000) as they have residents (Roberts 1995, p. 242). More than 80 percent of the 532 banks registered there maintain no physical branches other than a brass or plastic name plate in the lobby of another bank or law firm. United States banks have about 31.5 percent of their overseas assets in the Bahamas and Cayman. The money transferred there never comes physically. It is just booked in a ledger onshore (Roberts 1995, p. 244).

The Grand Cayman Islands are not the only tax haven in the world. Experts identify 41 countries and regions as tax havens. Together, these tax havens have a total population of 30 million and produce just 3 percent of the Western economies' Gross Domestic Product (Hines and Rice 1994, p. 150). The production reaches 3 percent only because of the inclusion of Switzerland and Hong Kong, which have substantial economic structures. Based on 1982 data from the United States Department of Commerce, tax havens accounted for $359 billion of the $1.35 trillion of corporate activity conducted worldwide by overseas affiliates of United States firms. The net income of tax havens represented $11.1 billion of United States firms' total foreign-source income of $36.0 billion (Hines and Rice 1994, p. 151).

The rise of the tax havens is a testimony to the herculean efforts of the tax avoidance industry. The people engaged in this industry may work hard; they may be dedicated to their job; they may even be brilliant in the execution of their responsibilities, but, to the extent that they are merely trying to bend the rules of the tax system, society gains nothing whatsoever from their effort.

Every few years, Congress announces its intention to simplify the tax code, but inevitably the result is still greater complexity and even more inequity. Building on the popular resentment of the tax code, some politicians began to advocate a flat tax. While the flat tax does provide simplification, it does so at the expense of fairness. Even the most prominent academic advocates of the flat tax admit that "it is an obvious mathematical law that lower taxes on the successful will have to be made up by higher taxes on average people" and that the plan "will be a tremendous boon to the economic elite" (Hall and Rabushka 1983, pp. 58 and 67). For example, the proposed flat tax falls on wages and salaries, while capital gains escape untouched.

I may be wrong, but despite its obvious advantages for the rich and powerful, I suspect that the flat tax will not be adopted in the near future. The present tax system is too attractive for too many of the major players.

It offers an ideal means of managing the economy while remaining hidden from the public view.

While Congress loudly debates a few elements of the tax code in order to appeal to various spectra of the electorate, I doubt if any elected representative has ever read, or even would be capable of comprehending the entire tax code. Once Congress gets ready to produce the final draft of a tax law, members slyly insert obscure phrases or paragraphs at the last minute in order to parcel out favors to political supporters. Frequently, these measures are worth many millions of dollars. With a flat tax, members of Congress would lose this opportunity to court the favor of potential contributors or future employers.

In any case, the wasted effort associated with the tax code is a dead-weight loss from virtually any perspective. In describing the tax system as a dead-weight loss, I am not enlisting in the current *jihad* against the government. My intent here is to begin an examination of the myriad of ways in which we dissipate our social potential.

War and the Reorganization of Society

Once we leave the realm of taxes, I would expect to find considerably more disagreement about the nature of economic waste. Rather than meet the reader's objections head on, let me continue to search for some common ground. Specifically, let us turn our attention to the way society treats efficiency during times of war.

Although war is perhaps the greatest waste ever invented by human beings, wartime pressures create a premium on other forms of efficiency. Indeed, during times of war, when the threat to social survival becomes extreme, countries abruptly abandon the rhetoric of individualism. Instead, we find a thoroughgoing emphasis on efficiency, except for the question of the inefficiency of the war itself. In the words of Ralph Hawtrey, a British Treasury official who was also a very influential economist:

> The War suddenly showed up all our economic standards from a new angle. In the belligerent countries people became aware of the paramount claims of the State. They were called upon to give up former pursuits in favour of a single transcendent purpose. The economic problem presented itself in an unequivocal form; human nature had to be worked upon and induced to do what the State required to do. Every combination of payment, persuasion and pressure was resorted to, not only to make people serve in the forces and work at the manufacture of munitions, but to regulate every part of their

lives. Controls and rationing were gradually extended in all directions. Markets ceased to function. Prices lost their ordinary significance. Demand usually meant either the demand of the State or a consumers' demand limited by rationing. Cost meant a total of wages and prices determined by authority and cut off from any semblance of free competition. Wealth no longer had any useful meaning. (Hawtrey 1925, p. 384)

For example, in the United States, a push toward standardization was an important part of the war effort during World War I. In forcing business to reduce the variation in styles and sizes, the government was "giving production problems precedence over sales problems" (Knoedler 1997, p. 1015; citing Haber 1964, p. 120).

The results of the standardization drive were impressive. The National Industrial Conference Board calculated that the War Industries Board's push toward standardization saved about 15 percent of total costs for the relevant industries (Knoedler 1997, p. 1015; citing National Industrial Conference Board 1929, p. 9; see also Knoedler and Mayhew 1994).

The standardization drive was part of a larger process. During World War II, John Maurice Clark observed that in peacetime society acts as if consumption were the objective; in war, leaders concern themselves with "the necessities of health, efficiency, and 'morale.'" In peacetime, we are limited by our capacity to consume; in war, by the capacity to produce (Clark 1942, p. 3). Earlier, during the World War I, Clark gave a slightly different interpretation of the changes that occur during war: "The need of a more coherent social organization is probably not less great in times of nominal peace, merely less obvious and less immediate" (Clark 1917, p. 772).

In other words, facing a serious threat of a military disaster, the typical individualistic motivations cease to occupy society. During such times, just as in the sensory deprivation chamber, we suddenly become aware of the need to eliminate wasted effort. Political leaders quickly realize that productive capacity depends upon both the capacity and the morale of the people.

For example, during World War I, the British *Report of the Ministry of National Service* told the country that only one man in three of nearly two-and-a-half million examined was completely fit for military service (Titmuss 1958, p. 81). Still later, during World War II, professor Cyril Falls said that, in military terms, the war could not be won unless millions of ordinary people, in Britain and overseas, were convinced that Britain had something better to offer than her enemies—not only during but after the war (Falls 1941, p. 13; cited in Titmuss 1958, p. 82).

Richard Titmuss, a British professor of social administration, noted that Family Allowances, the Beveridge Report, National Insurance (income security), and the Education Act of 1944 were all spawned during this time (Titmuss 1958, p. 84). He concluded: "The social measures that were developed during the war centred round the primary needs of the whole population irrespective of class, creed or military category. The distinction of privileges, accorded to those in uniform in previous wars, were greatly diminished" (Titmuss 1958, p. 82). In effect then, the pressing demands of a wartime emergency tend to encourage a nation to rationalize capital, labor, and society as a whole. In terms of rationalizing capital, the government takes measures to coordinate different industries and to see that firms use their plant and equipment as efficiently as possible. In terms of rationalizing labor, the government tries to make sure that the workers and soldiers are stronger and healthier.

The rationalization of society is the most subtle of the three rationalizations. Toward this end, the government tries to minimize the dissipation of effort wasted in conflict between various groups of people. For example, the government will act to minimize the disparities in the privileges that different classes enjoy.

Immediately after wars, before the memory of the threat has fully dissipated, political leaders still take an interest in policies that facilitate the creation of social solidarity and make the working class healthier. In this regard, Richard Titmuss noted: "It was the South African War, not one of the notable wars in human history, to change the affairs of man, that touched off the personal health movement which led eventually to the National Health Service in 1948" (Titmuss 1958, p. 80). In the United States, the School Lunch Program, established in 1946, was another classic case. Much of the initial support for the program was due to the persuasive testimony of Major General Lewis Hershey, Director of the Selective Service Commission. The general told congressional committees that, during World War II, poor nutrition accounted for many of the rejections of young men by local draft boards (U.S. House of Representatives 1989, p. 53).

Hawtrey noted that many hoped that the changes in society would go farther:

[culminating in] changes in human nature, which will bring new motives to bear. Mr. [Richard] Tawney looks forward to an extension through all occupations of the honorable zeal which we count on finding in the professions. This is itself a separate solution of the economic problem, a solution based like that of primitive society, upon a sense of obligation in the

> individual, but differing from the primitive solution in that the sense of
> obligation would be rational. It would take the form of a desire to render
> a service to society; it would not be bound up with a caste-imposed oblig-
> ation to render a service of a narrowly traditional kind, but would be free
> to adapt itself to the changing needs of society. (Hawtrey 1925, p. 385)

Of course, those in power naturally did whatever they could to forestall
such changes. In fact, even in the press of military emergencies, even when
leaders recognize the importance of encouraging social solidarity, the powers that be still have limits to how far they are willing to go in the direction of equality. Titmuss mentioned a particularly revealing example. In
May 1855 in the midst of the Crimean War, when Florence Nightingale
opened a reading room for injured soldiers in Scutari, the War Office responded that soldiers "would get above themselves" if, instead of drinking,
they read books and papers, and that army discipline would thereby be en-
dangered (Titmuss 1958, p. 85; citing Woodham-Smith 1951, p. 239, although this page reference seems to be wrong).

Lessons from Germany on Inequality and War

A recent collection of essays, comparing the experience of London, Paris,
and Berlin during World War I, suggests that, at least in part, Germany lost
that war because the German government was less able than either France
or Britain to persuade its people that it was acting fairly (see Winter and
Robert 1997; especially, Bonzon 1997, p. 302; and Triebel 1997). For example, Thierry Bonzon and Belinda Davis wrote that it was:

> [T]he unequal distribution of deprivation more than the deprivation itself
> that annoyed people the most. In all three cities the feeling of unequal ac-
> cess to food, of a growing gap between the excluded majority and a privi-
> leged few set a limit on the acceptance of sacrifices endured by individuals,
> families, and social groups for the sake of victory. . . .

Despite social tensions in London and Paris which should not be ignored
(and which largely contributed to the 'mobilization' of public powers on
these questions), these two objectives were met. In Berlin both were un-
reachable. . . .

The development of the black market in Berlin was no doubt the most
visible symbol of the contrast between lived experience in the German
capital and that on the other side of the line. Corruption existed every-

where, but only in Berlin did it emerge into a way of life, highlighting the extreme inequality of access to food in the German capital. . . .

More than the blockade or the successive bad harvests, the disorganization of the market and modes of distribution (blamed perhaps unfairly on shopkeepers and middlemen), the unequal distribution of essential foods within Berlin society, the link between the access for the most fortunate to the black market and the exorbitant price paid by the majority to obtain no more than reduced rations, all these fuelled public anger and public demand for urgent action by the state. (Bonzon and Davis 1997, pp. 340–41)

Gabriel Kolko, a renowned professor of history at York University in Toronto, noted that German workers did not recover their 1913 level of wages until 1928. The Nazis realized that they could not pursue their program of military conquest, if they repeated the mistakes of World War I and undermined social solidarity by intensifying inequality. In Kolko's words: "Forced to choose, the Nazis . . . preferred to risk depriving the war effort to possibly alienating the workers and seeing them driven once again to political action in various forms, including slowdowns and sabotage." So, in World War II, Hitler protected wages. As a result, "Real weekly income in Germany grew dramatically from 1932 to 1941, and even in 1944 it was only slightly less than it had been at its peak 1941" (Kolko 1990, p. xvii).

Herbert Hoover and the Elimination of Waste

Of course, during wartime emergencies, governments do not merely push egalitarian programs. They are also likely to abuse personal liberty. Such was the case when the United States rounded up Japanese Americans and imprisoned them in camps. Cynics who are sympathetic to laissez-faire might even suggest some sort of necessary association between the powers of the welfare state and a general limitation of freedom.

I raise the issue of wartime economies merely to indicate that when productivity rather than profit becomes a greater concern, societies are likely to restrain markets and turn to some type of rational planning. Even after a victorious war, the prestige of planning will carry over into peacetime.

World War I was no exception in this regard. For example, the call for standardization continued after the war. An industry-wide standardization effort by the Society of Automotive Engineers achieved savings of around $750 million, amounting to about 15 percent of the retail value of automobiles during the second decade of the century (Knoedler 1997, p. 1015; citing Thompson 1954).

The quest for efficiency went far beyond a call for standardization. Herbert Hoover was only one of a number of prominent engineers who "undertook to apply their scientific methods to what were seen to be the broader economic problems of the day—unemployment, overproduction, low wages, and industrial waste" (Knoedler 1997, p. 1012). Hoover estimated that American manufacturers could save $600,000 annually by standardization and the elimination of waste (Alford 1929, p. 119).

For example, the Federated American Engineering Societies, directed by Herbert Hoover, undertook the first comprehensive national study of industrial waste (American Engineering Council. 1921; see also Barber 1985, p. 7). This study proposed to examine the economy as a "single industrial organism and to examine its efficiency toward its only real objective—the maximum production" (Knoedler 1997, p. 1015; citing Layton 1971, p. 190).

Herbert Hoover was perhaps the most famous engineer of his day, well before he tired of making money and turned to the sphere of politics. As an engineer and a businessman, Hoover was enormously concerned about waste. This interest continued after he became a political figure. At the time of the publication of the study on national waste, he was Secretary of Commerce. Later, Herbert Hoover became the first engineer elected as president.

Even before his election, Hoover was not just an engineer, but a major political force. He used his position as Secretary of Commerce to encourage business to pursue simplification as well as standardization. He directed the Bureau of Standards to "set up new divisions to promote the adoption of commercial standards and simplified practices" (Noble 1984, p. 81). Working in cooperation with the United States Chamber of Commerce, Hoover established an operating procedure for simplification of products; it involved a thorough study of the particular problem area, followed by a meeting of representatives of manufacturers, distributors, and consumers to secure agreement to the elimination of certain types and sizes of products (Noble 1984, p. 81).

The new Division of Simplified Practices ultimately became the "medium through which producers, distributors, and consumers could agree upon simplification of production by reducing the number of sizes and models of products" (Noble 1984, p. 81). In his review of technical changes in manufacturing industries for the Hoover Commission study of *Recent Economic Changes,* he happily estimated that the Division of Simplified Practices had achieved a 98 percent reduction of varieties in some areas (Noble 1984, p. 82).

Herbert Hoover and the
Transcendence of Markets

Herbert Hoover believed that engineers could do far more than merely rationalize the existing economy (Barber 1985, p. 14). Toward this end, he wanted to go well beyond industrial measures to improve efficiency. For example, as Secretary of Commerce, he attributed a 10 percent increase in efficiency to the prohibition of alcohol (Fisher 1930, p. 173).

Hoover believed that engineers could greatly improve upon markets in other ways besides prohibiting the consumption of alcohol. As head of the newly created American Engineering Council, Herbert Hoover called for the "abandonment of the unrestricted capitalism of Adam Smith. . . . , for a new economic system based neither on the capitalism of Adam Smith nor upon the socialism of Karl Marx," but on the cooperation of all social groups (Wilson 1975, p. 42). Elsewhere, Hoover proclaimed that laissez-faire had been "dead in America for generations," except in the recalcitrant hearts of some "reactionary souls" (cited in Wilson 1975, p. 43).

Hoover set the tone for his approach in his foreword to the American Engineering Council study on waste, writing:

> The wastes of unemployment during depressions; from speculation and over-production in booms; from labor turnover; from labor conflicts; from intermittent failure of transportation of supplies of fuel and power; from excessive seasonal operation; from lack of standardization; from loss in our processes and materials—all combine to represent a huge deduction from the goods and services that we might all enjoy if we do a better job of it. (Hoover 1921, p. ix)

The specific examples Hoover gave showed that the problem was not merely a matter of technical efficiency; instead, the engineers had to take into consideration the way that society was organized. For example, the study itself made the obvious assertion: "Standardization of machine sizes would make possible the use of one machine for a greater variety of different jobs." Then it went on to explain what this meant in a market economy:

> [One company] paid $17,000 for a special press for printing a trading stamp. On losing this job, the press was scrapped, and later sold for $2,000. The contract in the meantime had been awarded to three other printers in succession, and each in turn had purchased a new press which he had to scrap or use disadvantageously at the expiration of his contract. (American Engineering Council 1921, p. 18)

Hoover's solution for the failures of competition was to let corporations organize markets, set standards and eliminate waste on their own, presumably led by their engineering staff (see Hawley 1981). Ironically, the Great Depression left Herbert Hoover, perhaps the most ambitious social engineer ever to hold the office of President of the United States, with the reputation of being an unthinking and doctrinaire ideologue of laissez-faire.

Thorstein Veblen and the Elimination of Waste

Thorstein Veblen, who worked briefly for Hoover's War Food Administration until he was fired, may have been the most creative United States economist who ever lived (Barber 1985, p. 200). His most popular book, *The Theory of the Leisure Class* (1899), drew attention to the waste associated with conspicuous consumption. He did so with a panache that might be unequaled among economists. Listen to a few of Veblen's own words:

> In order to gain and to hold the esteem of men, wealth must be put in evidence, for esteem is awarded only on evidence.
>
> This evidence consists of unduly costly goods that fall into accredited canons of conspicuous consumption, the effect of which is to hold the consumer up to a standard of expensiveness and wastefulness in his consumption of goods and his employment of time and effort. (Veblen 1899, pp. 42 and 88)

Veblen enjoyed comparing the lavish displays of wealth in his day with the seemingly primitive customs of the indigenous populations of the northwest region of North America, who were said to compete by destroying their wealth in potlatches, a ceremony in which the people were supposed to gain status by destroying their wealth. In effect, Veblen asked who was really the most primitive, the indigenous people who burnt their wealth in a public display or the luminaries of supposedly modern society who behave so foolishly in displaying their wealth.

Veblen applied his ideas of efficiency and practicality to his own life. He made nearly all his furniture himself out of dry-goods boxes, which he covered with burlap. The "chairs" were hard and they lacked backs, "but they were not unsuited to a hardy product of pioneer conditions" (Dorfman 1940, p. 305).

Veblen estimated that if household duties were properly organized they would not take more than an hour a day. He considered the ceremony of making beds to be a useless expenditure of energy. He preferred that the covers were merely turned down over the foot so that they could be easily drawn up at night. He washed dishes in an equally utilitarian fashion. He stacked them in a tub, and when all of them had been used, the hose was turned on (Dorfman 1940, p. 306).

Even before World War I, Veblen thought that standardization was a natural outcome of a rational industrial system. He proclaimed:

> Modern industry has little use for, and can make little use for, what does not conform to the standard. What is not competently standardized calls for too much craftsmanlike skill, reflection, and individual elaboration, and is therefore not available for economical use. (Veblen 1904, pp. 10–11)

Veblen was convinced that a market economy could not achieve anything like its maximum potential efficiency. In fact, Veblen considered business to be the very antithesis of efficiency. He considered that sabotage, which he defined as the "conscientious withdrawal of efficiency," was the defining characteristic of business (Veblen 1921, p. 38). He charged that "manoeuvres of restriction, delay, and hindrance have a large share in the ordinary conduct of business" (Veblen 1921, p. 3). If we looked at our economy objectively, he thought that we would realize that engineers rather than business people should be running it. In Veblen's words:

> In more than one respect the industrial system of today is notably different from anything that has gone before. It is eminently a system, self-balanced and comprehensive: and it is a system of interlocking mechanical processes, rather than of skillful manipulation. It is mechanical rather than manual. . . . For all these reasons it lends itself to systematic control under the direction of industrial experts, skilled technologists, who may be called "production engineers," for want of a better term. (Veblen 1921, p. 52)

The Elimination of the Retail Trade

Veblen wanted rational control to extend beyond the factories. Appalled by the wasteful duplication within the retail system, Veblen once made a proposal to scrap much of the distribution system of the United States and let the United States Post Office service the economy as a giant catalogue store (Veblen 1934, pp. 302–03). His idea was to combine the Parcel Post, the Post Office, and the Catalogue Stores into one great public entity.

Obviously, his proposed creation could not replace the entire retail sector. After all, fresh vegetables would not mail very well. Even so, just think of the savings that could occur if we made a move in the direction that Veblen proposed.

A Veblenized retail system would free up the time of millions of sales people. In addition, society could convert space from shopping centers to more productive uses. To put this possibility into perspective, the United States devotes more than four billion square feet of our total land area, or about 16 square feet for every American man, woman, and child, to shopping centers (Schor 1991, p. 107). Veblen's proposal would eliminate much of the resources used to construct and service our incredibly bloated retail structure, as well as a considerable amount of traffic congestion.

Just how practical was Veblen's proposal? Who knows all the ramifications? Today, we can see parts of the retail trade evolving in the direction of Veblen's vision, with the rise of commerce on the web. In addition, the success of some of the all-purpose warehouse stores suggests that Veblen may have been onto something important.

Now, many people enjoy roaming the malls, looking at the goods on display. In addition, our society is so alienating that for many people, shopping is a major source of social contact. However, many other people would appreciate converting our retail stores to parks and other public spaces.

Of course, if government legislated a Veblenized retail system overnight, the immediate effect would be widespread unemployment. In an irrational society, rational choices are often irrational. So, for the time being, think of Veblen's proposal as nothing more than a means of getting a handle on one dimension of the wasteful nature of our economy and return to the subject of standardization.

A Brief, but Contentious Note on Class and Standardization

Not everybody applauded the engineers' drive toward standardization. For example, while Hoover was pushing for standardization as Secretary of Commerce in the government of the United States, John Maynard Keynes, perhaps the greatest economist of the twentieth century, decried standardization. He complained that in the United States the number of types of glass bottles fell from 210 to 20, cigars from 150 to 6, collars from 150 to 25. He attributed this trend to advertising, lamenting that standardization has raised "the economic price of idiosyncracy" (Skidelsky 1992, p. 240; citing an unpublished note dated December 23, 1925).

Tibor Scitovsky, perhaps the only modern economist to take note of "the economic price of idiosyncracy," concluded:

> A person finds his tastes well catered to if he is conformist enough to share them with millions of others, because the things he then wants are profitable to mass produce and to sell at prices lowered by mass production. By contrast, a person with eccentric tastes who wishes to pursue a divergent lifestyle will be discouraged by the high prices or unavailability of the things he wants. (Scitovsky 1976, p. 10)

Where Hoover and the engineers saw the potential for great progress, Keynes and his elite friends mourned the loss of their ability to celebrate their uniqueness by purchasing goods that set them apart from their fellow citizens. About the same time that Keynes wrote his unpublished note, Victor Selden Clark must have been working on his massive, three volume *History of Manufactures in the United States.* Buried within the vast collection of details about technology, Clark insightfully observed: "Handicrafts and methods of production that follow the precedent of handicrafts, serve best an aristocracy of consumers, while factories serve best the consumption of a democracy" (Clark 1929, i, p. 528). In a similar vein, Keynes' close friend and biographer, Roy Harrod, later distinguished between "oligarchic wealth and democratic wealth" (Harrod 1958, p. 209; cited in Hirsch 1976, p. 23–24).

This concern about the nature of standardization might have been in the air during the 1920s, possibly, because the recent rise of mass production first made people sensitive to the possibility of a passing of the world of handicrafts. In the same year that Clark's book appeared, Allyn Young wrote that handicrafts exist in part "because of the insistent demand of buyers for special and particular characteristics in the things which they purchase" (Young 1990, p. 164).

Keynes's protests against standardization remind us that we cannot ever hope to get an accurate handle on the nature of efficiency. To some extent, efficiency, like beauty, remains in the eye of the beholder. Even apparent imperfections of workmanship in a handicraft product may give the wealthy the opportunity to distinguish themselves from the less fortunate.

While we might snicker at Keynes for being troubled by the lack of variety of bottles, the same drive toward standardization has more far-reaching effects in the realm of culture. For example, the publishing industry as well as the system for the distribution of books has been pruning away their selection of books. Once-proud publishing houses now satisfy themselves with spewing out so-called blockbuster books, by or about celebrities, while serious works have trouble reaching the press. This tendency toward sameness repeats itself in the cinema and in television. Perhaps the rush to standardization is most extreme in radio. One can scan the entire AM or FM spectrum in vain, hoping to find a single program that falls outside of the five or six standard formats.

In conclusion, recall that we initially turned to wartime conditions because of the tendency for more agreement about technical efficiency during such emergencies. Although the ideas of the wartime planners carried over into the postwar economy, so too did the prewar ideas about consumption that Veblen had so delightfully mocked.

We can see the confusion in the role reversals among Keynes, Hoover, and Veblen soon after the Great Depression had taken hold. The economic planner, Herbert Hoover, became the discredited icon of laissez-faire, while the sybaritic Keynes became the symbol of rational planning. All the while, Veblen, who died just before the Depression, is primarily remembered today as a satirist rather than a serious economist (Dorfman 1940, p. 508).

A Veblenesque Note on the Cost of Changes in Fashion

A good deal of our consumption is what Fred Hirsch, following Veblen, described as positional; in other words, consumption that is supposed to signal our status to others (Hirsch 1976). Enhancement of positional consumption does little to make our society better off. I can only climb up a rung on the positional ladder by ensuring that someone else declines. In this sense, positional consumption is a zero sum game that leads to no gain at all for society.

In another sense, positional consumption is a negative-sum game fueled by the profit motive. Hemlines rise and fall in order to make people dissatisfied with last year's wardrobe. Fashions change so fast that secondhand stores, such as the Goodwill or the Salvation Army, get far more donated clothing than they could possibly sell to their customers, even though much of it is high quality and relatively new. As a result, they end up exporting much of their donated clothing.

In one famous study, published in the conservative *Journal of Political Economy,* three prominent economists, Franklin Fisher, Zvi Griliches, and Carl Kaysen, estimated that more than 25 percent of the selling price of a car came from the cost of model changes that were unrelated to performance (1962). Since 1962, the speed with which new models of consumer goods proliferate has accelerated dramatically, in the automobile industry, but more so elsewhere. Jeffry Madrick recently wrote: "The number of new food and household products introduced each year, for example, has increased fifteen- and twenty-fold since 1970. The Gap retail chain revamps its product line every six weeks, and changes its advertising frequently as well" (Madrick 1998, p. 32).

The Swiss company that manufactures Swatch watches creates 140 different watch styles each year. I doubt a new model watch is much more accurate than the model that preceded it. The difference is in appearance rather than the ability to keep a more accurate measure of time.

Nike creates 250 new shoe designs each season (Jenkins 1998). While the ultimate purpose of product diversification is to increase profits, the company no doubt imagines that it is doing a service to its customers. In fact, diversification creates another form of waste, in a form that economists describe as search efforts. Writing as an aging basketball player with tender feet, I know that if I find a pair of shoes that fits well, I will never be able to find a replacement with the precise feel and fit, since the style that I buy today will soon be discontinued.

The continual need to change fashions reflects an impersonal world in which people communicate their status by displaying fashionable possessions. Since others will always attempt to ape the fashionable, the key to status in this game is the ability to discard old fashions and adopt new ones ahead of the masses. Veblen made his reputation comparing this competition to the potlatches of the Native Americans of the Northwest.

Business preys on this human frailty with advertisements demonstrating that success flows to those who follow the latest fashion, while ridicule and humiliation lies in store for those who either cannot or will not consume appropriately. In this environment, people who are uncertain of their position in the world stand in terror of being out of fashion—creating a "fashistic" world in which frightened consumers discard styles at breakneck speed.

In his book, *Everyday Life in the Modern World,* the French sociologist, Henri Lefebvre, included a chapter, "Terrorism and Everyday Life." He observed: "Not that fashion alone and independently causes terror to reign, but it is an integral/integrated part of terrorist societies, and it does inspire a certain kind of terror, a certainty of terror" (Lefebvre 1971, p. 165).

To the extent that new styles merely make us dissatisfied with our existing possessions, planned obsolescence imposes a significant drag on our economic potential. The time and energy devoted to these makeovers of our commodities could be devoted to improving our lives in a meaningful way. A more sensible society might devote less energy to devising new styles of tennis shoes every few days, while allowing for more variety in cultural matters.

CHAPTER 3

More Obvious Waste

Losses Due to Crime

Now that we have stumbled on to the controversies regarding standardization, let us return to some categories of waste that will elicit general agreement. Crime is an obvious example.

Crime creates a staggering loss to society. The fight against crime creates enormous government bureaucracies. Between 1982 and 1993, the cost of the criminal justice system in the United States soared from $37.8 billion to $97.5 billion dollars. By 1993, the system employed 1.860 million people nationally (United States Department of Justice 1997, pp. 2 and 18).

As of 1996, 1.5 million people were in jails or prisons in the United States (United States Department of Justice 1997, p. 502). In California, prisons now claim a greater share of the state budget than higher education.

We must also add the private cost the public must bear to protect itself against crime. Excluding the value of individuals' time, private expenditures on locks alone in 1985 was an estimated $4.6 billion (Laband and Sophocleus 1992, p. 960). No doubt, that figure has increased considerably since that time. In addition, we need to take the cost of other types of more sophisticated security gear into account.

Most economic estimates of the cost of crime do not typically include the burden that victims of crime impose on the medical system, but that cost is real. We also would have to include the physical, emotional, and economic costs to the victims of crime. We should also consider the contributions that these victims might have made had they not been caught up in a crime. Finally, we might count the emotional costs to those who fear being the victims of crime, even though they may have escaped that fate and even if the extent of that fear might be irrational.

So far, I would expect a general agreement to my notion of waste. I cannot resist carrying this discussion further onto more contentious ground. For example, as crime increases, police become more militarized. They begin to take on a more antagonistic position against anybody whom they regard as suspicious. As a result, police often abuse their authority, creating still another class of indirect victims of the expansion of criminal activity.

Incarceration itself represents an enormous loss to society. In part, this waste exists by design. The experience of James Gilligan, the former head of the prison mental health service in Massachusetts, illustrates this assertion. After reviewing the various programs for preventing recidivism, he found that the most successful was the one that allowed inmates to receive a college degree while in prison. Several hundred prisoners in Massachusetts had completed at least a bachelor's degree while in prison over a five-year period, and not one of them had returned to prison for a new crime. Gilligan recalled:

> Immediately after I announced this finding in a public lecture at Harvard, and it made its way into the newspapers, our new governor, William Weld, who had not previously been aware that prison inmates could take college courses, gave a press conference on television in which he declared that Massachusetts should rescind that "privilege," or else the poor would start committing crimes in order to be sent to prison so they could get a free college education. (Gilligan 1998, p. B 9)

While prison industries employ some inmates, generally on tasks requiring little skill, many criminals are extremely talented people. Some of the young people engaged in the drug trade display business acumen that would make the typical MBA student envious. To the extent that crime appears to be the best outlet for their talents, society loses the opportunity to benefit from the gifts of these people. Criminal records also compromise the future of those who run afoul of the law, since many employers are skeptical about hiring convicted felons. As a result, society loses much of the potential value of these people, even after they have abandoned their life of crime.

The War on Drugs

Of course, we could not eliminate all the expenses of containing crime, even in an ideal society. Some people may always require institutionalization. After all, a few people do represent a serious threat to others or to themselves, yet the United States incarcerates a greater percentage of its people than any other advanced society. However, no less an authority than Milton Friedman suggests that we could eliminate much crime with the stroke of a pen observing:

> The first and most obvious [way to reduce the amount of crime] is to reduce the range of activities that are designated as illegal. Surely, one reason for the growth in crime is that the number of activities that are classified as such, has multiplied in recent decades. (Friedman 1997)

The famous war on drugs is a case in point. Dan Baum's wonderful book, *Smoke and Mirrors,* reports on his interviews with the founding fathers of this ill-conceived "war" (Baum 1996). Some of them freely admit that the war on drugs had little to do with either public health or safety. Instead, they saw the stereotypical drug user as either an antiwar activist or an urban black. Not without reason, neither group had much affection for the Nixon administration. Attacking these "enemies" seemed to be a tempting opportunity to further the political agenda of the party in power. In Baum's words:

> [In the 1968 primaries] Nixon looked at "his people" and found them quaking with rage and fear: not at Vietnam, but at the lawless wreckers of their own quiet lives—an unholy amalgam of stoned hippies, braless women, homicidal Negroes, larcenous junkies, and treasonous priests. Nixon's genius was in hammering these images together into a rhetorical sword. People steal, burn, and use drugs not because of "root causes," he said, but because they are bad people. They should be punished, not coddled. (Baum 1996, p. 12)

Ronald Reagan continued Nixon's legacy by expanding the range of social problems that could be blamed on drugs (Baum 1996, p. 150). For example, he cites a paper from the Reagan administration, entitled "White House Stop-Using-Drug Program—Why the Emphasis Is on Marijuana":

> While OSHA (Occupational Safety and Health Administration) was created (in itself, a result, in part, of political pressure in Washington by anti-Big Business activists) and gushing regulations having to do with workplace machines and procedures, corporations themselves began attacking a major part of the problem where it really was—in alcohol and drug use by employees. (Baum 1996, p. 188)

We can catch a glimpse of the social purpose of the drug laws in the unequal application of punishment. Perhaps the most notorious example is the difference in the way that the law treats crack cocaine and powdered cocaine. A gram of crack cocaine (stereotypically associated with urban black youth) draws a far harsher sentence than an equal amount of powdered cocaine (commonly thought to be a misguided recreation of the more affluent). Should the offender happen to be a prominent member of society, we can be relatively certain that punishment will be a promise to enter a private clinic rather than a penal institution.

As a result of this war on drugs, the percentage of federal prisoners incarcerated for drug related crimes has risen from 16 percent in 1970 to almost 60 percent in 1997 (United States Department of Justice 1997, Table 6.0009). Largely as a result of the drug wars, incarceration in the United States has skyrocketed.

Today, the United States, along with Russia, is the world leader in incarcerations. The 1995 rate of incarceration in Russia was 690 per 100,000 people while the rate for the United States was 600 per 100,000. Imprisonment rates in the United States are six to ten times that of other industrialized nations, except for Russia.

Robert Higgs, a scholar at the Independent Institute, has calculated that at current rates of imprisonment, federal and state prisons will house 285 of every 1,000 black men, 160 of every 1,000 Hispanic men, and 44 of every 1,000 white men at some point in their lives. Unless we turn the legal system around, matters will only get worse. Higgs calculated that if the total incarcerated population continues to grow by 7.3 percent annually, it would double about every ten years, while the total population expands at one percent annually, the prison population will overtake the total United States population in the 2080s (Higgs 1999). Higgs goes on to point out that all the growth of the federal inmate population since 1988 has resulted from the addition of drug offenders, yet drug usage hasn't budged.

Corporate Crime

While a tough-on-crime attitude is popular throughout the electorate of the United States, "crime in the suites," to use Ralph Nader's felicitous expression, escapes the harsh penalties meted out to those who perpetrate crime in the streets. In fact, our legal system is generally forgiving of white-collar crime.

When corporations do find themselves indicted for criminal behavior, the same sort of inequality that we find in the application of our drug laws replicates itself. Relatively few large corporations face indictment. Of the 1,283 corporations convicted of federal crimes from 1984 through 1987, only about 10 percent crossed the threshold of $1 million in sales and 50 employees; less than 3 percent had traded stock (Cohen 1989, p. 606).

Total monetary sanctions imposed on the small number of corporations that are convicted represent an average of only 33 percent of the monetary value of the harms caused. If we only include direct fines, then the ratio falls to a mere 10 percent (Cohen 1989, p. 618: Table 6).

In addition, the penalties for small offenses are relatively higher than for large ones. In fact, we can see a strong inverse correlation between the size of the harm done and the sanction as a percentage of the magnitude of that harm (Cohen 1989, pp. 617–68). For example: "a firm causing up to about $50,000 in harm can expect to pay about twice that amount in criminal penalties and restitution. However, a firm causing over $1 million in harm is likely to pay less than the harm it caused in criminal fines and restitution" (Cohen 1989, p. 658). In short, the petty criminal, whether in terms of the size of the corporation or the enormity of the crime, is likely to face higher sanctions. Knowing that the likelihood of a penalty is small and the that likelihood of a serious penalty even smaller, firms have little reason to fear the consequences of their actions.

We find a similar pattern when the government imposes penalties on military contractors for defrauding the government. One recent study found that the extent of losses of firms that are fined for such violations of the law vary greatly when measured by the loss in the value of their stock. For example, the average effect for one of the top 100 defense contractors is a statistically insignificant 0.39 percent of value of the firm. For firms that are not among the 100 largest, the mean effect is more than 10 times as large, at 4.46 percent (Karpoff, Lee, and Vendrzyk 1999).

One case seems to stand out as an exception. In perhaps the most famous case of its kind, in 1961, the government convicted Westinghouse and General Electric for price fixing. These companies, along with some lesser firms, met to conspire to rig their bids and then falsified records (Mokhiber 1988, p. 217). F. F. Loock, president of the Allen Bradley Company, one of the corporations indicted by the Philadelphia grand jury in the price-fixing conspiracy case, commented about a meeting the conspirators attended in order to set prices: "No one attending the gatherings was so stupid that he didn't know that the meetings were in violation of the law. But it is the only way a business can be run. It is free enterprise" (Mokhiber 1988, p. 218; citing Smith 1961, p. 178).

As might be expected, the judge imposed relatively trivial fines compared to the earnings of the giant corporations. General Electric, which faced the largest judgement, had to pay only $457,000, an amount that represented only one-tenth of 1 percent of its earning over the five-year period that the conviction covered (Herling 1962, p. 319).

As John Herling, the author of a book on the case, noted, General Electric was hardly a first time offender:

In December, 1961, the Justice Department . . . listed 39 antitrust actions against GE, 36 of them since 1941. These included 29 convictions, seven consent decrees, and three "adverse findings" of the Federal Trade Commission. To the Justice Department this indicated "General Electric's proclivity for persistent and frequent involvement in antitrust violations" in all branches of industrial production. Westinghouse could show almost as cluttered a record in antitrust violations. (Herling 1962, fn p. 320)

While the monetary dimensions of this judgement were insignificant, the penalty did have one unusual feature. The judge sentenced 3 vice presidents to spend 30 days in jail (Mokhiber 1988, p. 220). I recall the reactions of business people at the time who were shocked that respectable executives would find themselves in prison.

The "harsh" penalties paid by the three executives seemed to send a message to the corporations. The price of turbine generators and other heavy electrical equipment soon fell by nearly 70 percent (Keller 1977, p. 74). Alas, nothing similar has happened for four decades.

Capital's Capital Crimes

A "bleeding heart executive" might find the jail time too harsh a price for a fellow manager to pay. After all, price fixing may be a form of theft, but after all nobody suffered any direct physical harm from price fixing. In fact, other forms of corporate criminal negligence do result in an enormous amount of physical harm. A worker dies on the job from injury or harmful exposure every 11 minutes according to the National Safety Council (1998).

A very conservative estimate, published in *Archives of Internal Medicine,* a journal of the American Medical Association, puts the problem of workplace safety into perspective. Workplace injuries and illnesses cost an estimated $172 billion each year and result in approximately 6,500 deaths from injury and more than 60,000 deaths from disease. Workers suffer an estimated 13.2 million nonfatal injuries and 862,200 illnesses annually (Leigh, Markowitz, Fahs, Shin, and Landrigan 1997).

Another study observed that while 24,000 die each year as a result of crime in the streets, about 25,000 die from workplace exposure (Reiman 1996, pp. 65–66). However, since only about half the population works on a job site and everybody is a potential victim of street crime, we might say that a worker is twice as likely to die from workplace exposure as from common crime.

Of course, we cannot hold business responsible for all of these deaths and injuries. No doubt, in some cases the workers were responsible for their own demise, but certainly the employers bear the primary responsibility for the majority of these deaths. In addition, the number of deaths caused by corporations must include the tragedies experienced by those who die from consuming dangerous or defective products. Cigarettes are, perhaps, the most obvious example. In addition, others die from enormous amounts of toxic materials that businesses emit.

Some of the harm that business causes is perfectly legal. Cigarettes are a case in point. However, illegal business practices lead to many of the deaths and injuries. How seriously does the government regard corporate crime? In 1973, 1,500 federal marshals guarded airplanes from hijackers, while the Occupational Safety and Health Administration had only 500 inspectors to check on all of the workplaces of the United States (Reiman 1996, p. 70). A 1992 AFL-CIO report put the protection of human life into perspective: First, the government spends $1.1 billion per year to protect fish and wildlife compared to $300 million to protect workers from health and safety hazards on the job (Reiman 1996, p. 70). Similarly, in the wake of a 1991 fire at Imperial Foods' poultry plant in Hamlet, North Carolina, where 25 workers lost their lives in another fire, the state's commissioner of labor, James Brooks, admitted, "North Carolina has more people on the governor's personal security force than protecting the health and safety of 4 million workers at 180,000 workplaces" (Dorman 1996, p. 5).

Although business portrays the cost of complying with regulations to protect workers as a waste, the need for workplace protection is enormous. In 1985 the National Center for Health Statistics of the U.S. Department of Health and Human Services asked a sample of over 100,000 workers to assess the safety of their own jobs. An astounding 33.9 percent reported working with hazardous substances; 35.2 percent reported hazardous conditions; and 39.3 percent reported risks of injury, even though the survey included officials, administrators, personnel, and sales workers as well as production workers (Dorman 1996, p. 12).

Since then matters have become even worse. Statistics collected by the National Institute for Occupational Safety and Health demonstrate the long-term historical trend in injuries for four major industries with significant safety problems. In each case the record is U-shaped, with a historic high earlier in the century, steady reduction leading to a low generally in the 1960s or early 1970s, and then an upturn heading into the 1980s (Dorman 1996, p. 14). After the obligatory dip for the 1982 recession, the number of lost workdays per employee due to occupational injuries or illnesses virtually explodes: the increase is more than 30 percent (Dorman 1996, p. 15).

Although workplace injuries represent a serious form of waste, this particular waste weighs lightly in public debates because it imposes relatively few costs on business. Indeed, direct oversight of health and safety conditions would not be quite so vital if firms had to pay a substantial penalty in the event of a mishap. Unfortunately, during the 1972–1990 period, an incident resulting in death or serious injury resulted in a median penalty of a paltry $480 (Reiman 1996, p. 70).

Surprisingly, even these minimal penalties seem to have an effect. One study estimated that, based on a Bureau of Labor Statistics panel of inspections of 6,842 manufacturing plants for 1979–1985, if an inspection results in a penalty, the plant experiences a 22 percent decline in injuries during the next few years (Gray and Scholz 1991). I presume that if the prevention of injuries were very expensive the firms might be less likely to take action, given the low penalties imposed.

Although the typical worker stands a significantly greater chance of being harmed from the work environment than from what we consider to be street crime, the penalties for creating the dangerous conditions on the job are virtually nonexistent.

Presently, the media and the politicians have whipped the public up into a frenzy about street crime. All the while, capital's crimes pass by virtually unnoticed.

The Costs of the Political System

Politicians are not likely to raise the issue of corporate crime, since they are so beholden to the corporations for their campaign contributions. This behavior comes as no surprise. After all, we are all familiar with the undemocratic abuses associated with the purchase of politicians' services through donations to political parties or campaigns. Let us look at this system in terms of its economic costs.

Economists who look more favorably upon the market than upon public activities emphasize the waste associated with what they call, "rent-seeking activities." The term, "rent-seeking" begins with an article by Anne Krueger (1974). Over time as antigovernment rhetoric began to heat up, the reported estimates of the losses from rent-seeking began to jump by leaps and bounds. For example, Kreuger estimated the annual welfare costs of rent-seeking induced by price and quantity controls to be 7 percent of the gross national product in India and a somewhat higher percentage in Turkey. Within a decade, a new study estimated that the costs of rent-seeking in India were on the order of 30–45 percent of GNP (Mohammad and Whalley 1984; cited in Anderson, Goeree, and Holt 1998). Soon thereafter, Richard Posner, now a federal judge, estimated the social cost of regulation to be up to 30 percent of sales in some industries (motor carriers, oil, and physicians' services) (Posner 1975; cited in Anderson, Goeree, and Holt 1998). Without going into detail about Posner's estimates, one need only consider the fate of medical care in the United States since the rise of the health maintenance organizations.

Three of the leading proponents of the rent-seeking approach, James Buchanan, Robert Tollison, and Gordon Tullock offer the definition: "rent-seeking is meant to describe the resource-wasting activities of individuals in seeking transfers of wealth through the aegis of the state" (Buchanan, Tollison, and Tullock 1980, p. ix).

The authors crafted this definition to be more narrow than it appears, since they assume that what the market does is the norm. For example, within the context of their definition, to petition the government to make polluters pay is rent-seeking, while calling for the government to diminish its regulatory activities is just sound economics.

I prefer a broader approach. The waste associated with this broader concept of rent-seeking reaches enormous proportions. Of course, the defects of the political process are so blatant that pointing out the egregious wastes associated with it is like shooting fish in a barrel. Nonetheless, let us take a moment to consider some of the most obvious wastes associated with the political process.

During the 1996 election cycle, business contributed $653 million dollars. Others contributed around $200 million, according to the Center for Responsive Politics. By the time you read this, the costs will no doubt have soared once again to new record levels.

The major financial cost of elections is the purchase of television time. Payment for television time is doubly insulting. To begin with, although the airwaves are, by any rational analysis, public property, a few corporations have been given exclusive use of this resource. Despite the incredible possibilities of television as a tool for communication, most political advertising, just as is the case with advertising in general, is used to confuse the electorate and spread disinformation. The (not necessarily unintended) result is that people become so cynical that only a few even bother to vote.

Besides the direct expense of influencing the electorate, political interests employ an army of lobbyists in a further attempt to control the political process. Kevin Phillips estimated that in 1991, about 91,000 lobbyists and people associated with lobbying worked in and around Washington. Phillips counted 63,000 people working for trade and membership associations in Washington in 1993. The number of U.S. corporations with offices in Washington grew from under 100 in 1950 to more than 500 in 1990, probably employing 5,000 to 7,000 people. Two-thirds of the fifty largest multinationals have offices in Washington. Close to 400 foreign corporations have some kind of representative on hand (Phillips 1994, p. 34).

An Associated Press report from July 8, 1998 calculated that lobbyists spent $1.7 billion to influence Congress in 1997, over and above what they spend to sway the administrative branch of government. Ken Silverstein caught the flavor of this abominable waste of resources in his rollicking *Washington on $10 Million a Day: How Lobbyists Plunder the Nation* (1998).

We also need to pay attention to the torrent of resources that flow into the public relations business. Public relations goes far beyond putting out press releases. Public relations firms attempt to create the appearance of a groundswell of support for their clients. Some of these firms also pay operatives to infiltrate organizations that are opposed to the interests of their clients (Stauber and Rampton 1995).

Of course, public support does not necessarily determine the outcome of political processes. For example, political decisions on major issues in the United States followed public opinion polls about 55 percent of the time between 1980 and 1993, a decline from the 63 percent shown in an earlier study of the 1960–1979 period (Monroe 1998); however, the less intense public opposition is, the easier it will be for a corporate interest to succeed in flaunting the will of the people.

Sadly, the direct wastes associated with the manipulation of the political system pale in comparison to the total losses. A corporation spends millions of dollars to buy influence with a politician only because they expect a return from that investment many times greater than their out-of-pocket costs. Such rent-seeking is truly profitable.

As a result of their efforts, wealthy manipulators of the political process enjoy outcomes that go against the best interest of the majority of the people, including a wide array of benefits, such as free use of public resources, freedom from regulation, tax giveaways, or lucrative contracts.

In any case, the waste associated with this system is outrageous. From time to time, public attention briefly turns to the bloated Pentagon budget, littered with unnecessary weaponry. Within a short time, patriotic fervor, together with imagined threats from enemies as feeble as North Korea, soon distract the body public, allowing the military to accumulate even more funds. All the while the enormous fiscal resources of the government neglect the real needs of the people.

Litigation

Litigation creates an enormous waste of resources. One study of 35 countries used the presence of physicians as an indicator of productive labor. They found that the more lawyers per physician, the slower growth was (Magee, Brock, and Young 1989, pp. 119–20).

A recent actuarial study by Tillinghast-Towers Perrin indicates that tort costs rose from $67 billion in 1984 to $152 billion in 1994 (United States Congress 1996; citing Tillinghast-Towers Perrin 1995). The press often gives the opinion that the majority of these cases involve product liability, but Ralph Nader's Public Citizen's Web site reports that this category accounts for only about 40,000 of about 19.7 million civil cases filed annually according to the National Center for State Courts (Public Citizen 1998).

Michael Rustad, a law professor who reviewed all the empirical studies of product liability cases, reported that each and every one of these studies concluded that punitive damages verdicts are rare (Rustad 1998, p. 54). In addition, he found that "[a] typical defendant in a products liability case is 'a national corporation, not a small business. The majority of primary defendants assessed punitive damages in medical malpractice cases are corporate defendants such as nursing homes, hospitals or health care organizations" (Rustad 1998, p. 54).

In a sense, I am surprised that product liability suits are not more frequent. For example, caveat emptor is an inadequate principal for regulating product safety. With the massive deregulation of business during the last two decades, the courts represent perhaps the last source of protection for the public. Only massive judgments will prevent business from cutting corners, even where the public safety is put at risk.

In reality, suits between corporations occupy far more of the court's time. The suit between Texaco and Penzoil is emblematic of the costs of such litigation. From 1984 to 1988, these two corporate giants battled in the courts. Penzoil initially won a judgment of more than $10 billion, but the ultimate settlement was a mere $3 billion. In the aftermath of the suit, Texaco's value fell by far more than Penzoil's rose because of the cost of the litigation. In fact, the combined equity of the two firms fell by about $21 billion, many times more than the legal fees incurred in the course of the litigation (Cutler and Summers 1988).

We are a litigious society because we are a divided society. A French economist, Phillipe Fontaine, upon reading an early version of this manuscript reminded me that French society is also divided, but not nearly as litigious as that of the United States. Within the business climate of the United States, corporations try to exploit the legal system to take advantage of other corporations, as well as the public. With the rise of the economic importance of ill-defined intellectual property, we can expect an explosion of litigation between corporations. The result of these suits is certain to be an enormous increase in the waste of resources associated with maneuvering for economic advantage in the courtroom.

Where people share values, litigation is less prevalent. For example, Robert Ellickson examined the effects of open- versus closed-range laws on cattle trespass disputes in Shasta County, California. Under open range laws, cattlemen are not usually responsible for accidental trespass damage, whereas they are strictly liable under closed-range laws. His point is that cattlemen and their neighbors cooperate to resolve their disputes regardless of who is liable.

Following Robert Frost's dictum that "Good fences make good neighbors" (Frost 1977), Ellickson suggested that, rather than litigate, individuals in Shasta County seem to rely on community norms to determine their behavior (Ellickson 1991). Living near to Shasta County, I believe that Ellickson overstated the degree to which cooperation occurs. Although Ellickson did describe a few contentious conflicts, he tends to downplay their importance since he favors less government regulation. Even so, Ellickson performed a valuable service by showing how widely-accepted social norms can help to resolve problems that might otherwise end up in acrimonious litigation, which can lead to further conflict.

CHAPTER 4

Social Conditions and the Absence of Trust

Classes

The root cause of most of the wastes that we have encountered so far is not technological. Instead, we must look to social conditions. In particular, we can trace much of the waste to the existence of a class structure that broadly divides people into workers and owners of capital. Consider the very existence of government, which many people consider to be the epitome of waste. Hunter-gatherers did not have governments for an obvious reason. After all, in the words of Jared Diamond: "Since . . . hunter-gatherers . . . did not produce crop surpluses available for redistribution or storage, they could not support and feed nonhunting craft specialists, armies, bureaucrats, and chiefs" (Diamond 1997, p. 55). Once early societies developed the capacity to produce more than their basic needs, one group managed to claim for itself a disproportionate share of the fruits of the earth, as well as the earth itself. Members of this group then excused themselves from the need to work by the sweat of their brow, depending, instead, upon the efforts of others. They enjoyed the status of what later ages termed unproductive labor.

This division of the world into classes created a need for an effective apparatus to protect the property and privileges of the well-to-do and to ensure that the less fortunate continue to labor with sufficient diligence to support the lifestyle of their betters. Naturally, not everybody was pleased with this arrangement.

Long ago, Adam Smith warned that the working classes were possessed by "passions which prompt [them] to invade property, passions much more steady in their operation, and much more universal in their influence"

(Smith 1776, V.i.b.2, p. 709). Smith noted that "in the poor the hatred of labour and the love of present ease and enjoyment, are the passions which prompt to invade property, passions much more steady in their operation, and more universal in their influence" (Smith 1776, V.i.f.50, pp. 781–82).

Consequently, Smith proposed that government is necessary to protect the property of the rich (Smith 1776, V.i.b., pp. 670ff). He even went so far as to teach his students:

> Laws and government may be considered in . . . every case as a combination of the rich to oppress the poor, and preserve to themselves the inequality of the goods which would otherwise be soon destroyed by the attacks of the poor, who if not hindered by the government would soon reduce the others to an equality with themselves by open violence. (Smith 1978, p. 208; see also p. 404)

Ultimately, however, protection of property depends upon the willingness of the broad mass of people to condone the existing distribution of property. In the words of John Stuart Mill, a mid-nineteenth-century British economist whom we will discuss later in this chapter, "Much of the security of person and property in modern nations is the effect of manners and opinion rather than of law" (Mill 1848, I.vii.6, p. 114).

Consequently, those who control the bulk of the property devote considerable effort to convince the rest of society that the status quo is in the best interest of all, rich and poor alike. Perhaps the culmination of this project was the creation of the theory of trickle-down economics. Unfortunately, inequality contributes mightily to the waste that plagues contemporary society.

Inequality and Hostility

Inequality breeds perverse psychological effects among the rich. As people have known for millennia, inequality fuels social discontent that goes well beyond the threat of criminal acts against property owners. In the words of Aristotle: "when men are more equal, they are more contented. But the rich, if the constitution gives them power, are apt to be insolent and avaricious" (Aristotle 1988, p. 122). We can even find echoes of Aristotle's insight in Adam Smith, who worried that such attitudes impair the functioning of the economy (Perelman 2000).

We can extend Aristotle's understanding to say that a relatively equal distribution of income creates a society where people can be more at ease with each other. Richard Tawney, who wrote what may have been the most eloquent plea for equality, observed: "If riches are an economic good, and are the proper object of economic effort, equality may, nevertheless, be a social good, and be made no less properly the object of social effort" (Tawney 1929, p. 33). Some writers have even proposed that we should see equality as a public amenity, such as roads, which improve the quality of our lives (Thurow 1971). While the amenities associated with equality are welcome, as an end in themselves, they also promote economic growth and well being.

The concern for equality goes beyond aesthetic considerations. Aside from the value of equality as a social good, we should keep in mind that inequality has serious economic consequences. As Tawney noted: "The hostility and suspicion resulting from inequality are themselves one cause of a low output of wealth" (Tawney 1929, p. 152). This hostility and suspicion in turn leads to political instability. Since business craves security, this environment discourages the very investment that lies at the heart of the justification of inequality that the trickle-down theorists justify. As a result, inequality generally reinforces the poverty of those at the bottom.

The poor are not the only people to feel this hostility and suspicion toward wealth. A recent *Wall Street Journal* article reported that as the job market surged and the stock market has boomed, a wave of envy began to gnaw at those near the top of the economic pyramid, as they saw others making even more. Among the most unsettled are successful corporate managers and professionals, such as doctors and lawyers, earning $100,000 to $200,000 a year. Even though these people are doing better, in terms of relative wealth, they are slipping. Since 1990, households earning more than $200,000 have seen their share of national income jump to 18 percent from 14 percent. In 1995, a household with an income of $288,000 ranked in the top 1 percent of the country. By next year, it will need to be making $385,000 to retain that ranking (Kaufman 1998).

Exploring the Economic Waste of Inequality

Traditional trickle-down economic theory is meant to console those who find themselves in the less privileged strata of class society. According to this theory, an unequal distribution of wealth should be favorable to growth. After all, the rich are supposedly more likely to save and to invest than the poor. The more resources that the rich have, the greater investment society will enjoy.

While the theoretical justifications of inequality might sound reasonable, trickle-down theory, like all economic propositions in general, rests on a fragile foundation of a few broad, but insupportable assumptions. In the process, the proponents of inequality sweep vital aspects of reality under the proverbial rug.

I will discuss why unequal societies actually grow more slowly than more egalitarian societies. This situation should not be surprising. Common sense tells us that once the overwhelming majority of a society comes to resent the undeserved privileges of the fortunate, society just does not work as well. We might also recall Adam Smith's discussion of "the passions which prompt [the poor] to invade property." Relatively few people realize today what seemed obvious to Adam Smith—that in a society of rampant inequality, the state must mobilize valuable resources in order to contain the anger and resentment of the poor.

Despite his measure of realism, Adam Smith was, deep down, very conservative. He attributed the problems of inequality to the irrationality of the poor; that is, their passions. Presumably, Smith would have us believe that a dispassionate analysis of the state of affairs in a highly unequal society would lead the poor to accept the wisdom of trickle-down economics that inequality can contribute to a healthy rate of growth.

In truth, we can recognize rational explanations for the poor performance of unequal societies. Take the example of crime. Where people see few legal opportunities open to them, they are more likely to turn to crime (Freeman 1996).

Since the mid-1970s, real wages paid to men 16–24 years old who work full time have fallen 20.3 percent. Real hourly wages paid to all-male hourly workers between 16 and 24 years old fell by 23.0 percent (Grogger 1998, p. 784). According to one study, the decline in wages for unskilled men from 1980 to 1994 explains up to an estimated 60 percent of the increase in property crime and only 8 percent of the increase in violent crime during that period (Gould, Weinberg, and Mustard 1998). Another study of arrests for a wider range of crimes reported even more dramatic results. It estimated that the response to a 20 percent fall in wages for young people should be an increase in youth participation in crime of 20 percent (Grogger 1998, p. 785). Indeed, between the early 1970s and the late 1980s, arrest rates for 16-to-24-year-old males rose from 44.6 to 52.6 per 1,000 population, a gain of 18 percent.

According to the United States Department of Justice, the average inmate was at the poverty level before entering jail. Almost half of jail inmates had reported incomes of less than $600 per month in the month before their most recent arrest (United States Department of Justice 1998, pp. 35 and 4).

In other words, inequality is responsible for a good deal of the crime that so troubles the public. International comparisons support the association between crime and inequality. The more unequal the distribution of income, the higher the incarceration rate (Doyle 1999).

Conservatives also accept the logic of crime as a rational calculation, but rely on harsher penalties to deter people from criminal activity. Ironically, conservatives fail to apply the same logic to the misdeeds of corporations. At least, they are silent about the lax enforcement of the law when the violator is a corporation rather than a poor young man.

In fact, the logic of punitive deterrence should be stronger for the corporate sector than for individual offenders. For poor, desperate people, with little to lose, the threat of punishment often does not act as much of a deterrent. Given the likely opportunities open to the disadvantaged, even a rational decision maker, cognizant of the possible consequences, might choose crime. In contrast, corporations, always sensitive to bottom-line considerations, would be likely to be deterred from stronger penalties.

The costs of inequality go far beyond the need for prisons and police. Over and above the problem of crime, inequality creates any number of problems.

I shall discuss how opportunities for trust and cooperation evaporate in an unequal society. In addition, an unequal society will enjoy poorer systems of education and health, both of which will reduce the economic potential.

Inequality and Growth

The study of Alesina and Perotti, covering a sample of 71 countries for the period 1960–1985 confirms the negative association between inequality and economic growth (Alesina and Perotti 1996).

Derek Bok, who should know about such matters from the vantage point of his previous experience as president of Harvard University, observed:

> The ultimate reason why we cannot ignore unjustified wealth is that it weakens the public's faith in the fairness of the economic system. Such faith is essential if we are to maintain support for the social order and inspire individuals to observe the laws, undertake the duties of citizenship, and extend the minimum of trust toward institutions necessary for communities to prosper. (Bok 1993, p. 231)

Samuel Bowles and Herbert Gintis, two prolific graduates of Harvard's doctoral program in economics, offered a few examples of how inequality makes society work less well, noting:

> Inequality fosters conflicts ranging from lack of trust in exchange relationships and incentive problems in the workplace to class warfare and regional clashes. These conflicts are costly to police. Also, they often preclude the cooperation needed for low-cost solutions to coordination problems. Since states in highly unequal societies are often incapable of or have little incentive to solve coordination problems, the result is not only the proliferation of market failures in the private economy, but a reduced capacity to attenuate these failures through public policy. (Bowles and Gintis 1995, p. 409)

Bowles and Gintis expanded their analysis of the costs of inequality even further, writing:

> Enforcement activities in the private sector may also be counted as costs of reproducing unequal institutions. Enforcement costs of inequality may thus take the form of high levels of expenditure on work supervision, security personnel, police, prison guards, and the like. Indeed, one might count unemployment itself as one of the enforcement costs of inequality, since the threat of job loss may be necessary to discipline labor in a low-wage economy. . . . In the United States in 1987, for example, the above categories of "guard labor" constituted over a quarter of the labor force, and the rate of growth of guard labor substantially outstripped the rate of growth of the labor force in the previous two decades. (Bowles and Gintis 1995, p. 410)

A host of recent studies has borne out Tawney's assertion that a more unequal distribution of income causes the economy to grow more slowly (Alesina and Rodrik 1994, p. 485; Alesina and Perotti 1996; Persson and Tabellini 1994). One study estimated that a reduction in inequality from one standard deviation above the mean to one standard deviation below the mean would increase the long-term growth rate by approximately 1.3 percent per annum; however, the measure may be biased toward zero. Using a different technique, the increase would be 2.5 percent (Clarke 1995, p. 423).

While a change of a little over a percentage point might not seem very significant, such differences in growth rates compounded over a couple of decades result in a gap in the level of incomes. For example, if in 1960 the Republic of Korea had Brazil's level of inequality, Korean per-capita income in 1985 would have been 15 percent lower, representing a loss of about two years' growth (Birdsall, Ross, and Sabot 1995, p. 496).

To my knowledge, the existence of this weak spot in the theory of trickle-down economics has not entered into public debates about inequality. Instead, the bulk of the economics profession echoes the virtues of trickle-down economics, adding that the inordinate rewards that the fortunate few can enjoy will spur the rest to emulate them.

For example, Finis Welch, who gave the prestigious Richard T. Ely lecture at the 1999 meeting of the American Economic Association entitled his talk, "In Defense of Inequality." Welch proclaimed:

> I believe inequality is an economic "good" that has received too much bad press. . . . Wages play many roles in our economy; along with time worked, they determine labor income, but they also signal relative scarcity and abundance, and with malleable skills, wages provide incentives to render the services that are most highly valued. . . . Increasing dispersion can offer increased opportunities for specialization and increased opportunities to mesh skills and activities. (Welch 1999, pp. 1 and 15)

Social Problems as Symptoms of Inequality

Earlier, I discussed the inordinate magnitude of the economic losses due to crime, especially to drugs. At the present time, cocaine seems to be doing more damage to society than most other drugs.

I think of cocaine as a particularly "economic" drug, in the sense that it seems to be most attractive to people on either extreme of the economic spectrum. While crack cocaine seems to have penetrated some of the poorest parts of our society, powdered cocaine seems to be especially welcome among some of the richest strata of our society.

Admittedly, we have very poor data about such matters, but if my suspicion is correct, then cocaine use might be good indicator of the degree of inequality in our society. I would go further and make the commonsensical suggestion that a reasonable person might expect a more egalitarian society to have less crime, fewer and less expensive measures to protect against crime, and a more modest criminal justice system.

I have already noted that where young people see no probable opportunity for advancement, they are more likely to turn to crime. In addition, study after study has shown that unemployment is strongly associated with crime. Even Aristotle recognized this problem, observing:

for it is activities exercised on particular objects that make the corresponding character. This is plain from the case of people training for any contest
or action; they practise the activity the whole time. Now not to know that
it is from the exercise of activities on particular objects that states of character are produced is the mark of a thoroughly senseless person. (Aristotle
1908, Book 3, Section 5, 4th Para., p. 61)

In addition, inequality is harmful because, as I mentioned before, unequal
societies tend to have less widespread education (Fernandez and Rogerson 1996). We hear much about the importance of education for our economic future, but as long as our society maintains the wide gulf between
rich and poor, we can expect that the electorate will refuse to mobilize
enough public funds to finance adequate education for the majority of
students. After all, relatively few poor people vote. For the very wealthy,
who have a disproportionate share in the electoral process, the cost of
sending a child to an expensive private school is trivial compared to the
cost of paying a fare share in a tax system that supports quality education
for all young people.

Inequality is also detrimental to good health. In fact, emerging research
indicates that inequality harms the health of rich as well as poor, although
certainly not to the same extent (Wilkinson 1997). At first, this association
might seem counterintuitive. However, unequal societies are more stress-
ful. Recall the earlier discussion about the need for more control in unequal societies. This control creates stresses for the controllers as well as the
controlled.

In addition, the islands of poverty offer ideal breeding grounds for diseases. People weakened by stresses associated with poverty are less able to
fight off diseases. They may also be more susceptible to dangerous behavior patterns that make them more prone to disease, such as the sharing of
needles. Oftentimes, diseases that arise in such conditions have more social
mobility than the people who carry them. As a result, the pathogens bred
in poverty can strike the wealthy as well.

I already mentioned that inequality tends to promote distrust. Next I
will turn to the harmful effects of this distrust.

On Leviathans, Lemons, and Trust

The losses from class conflict that Adam Smith described cover only a
small part of the dissipation of economic potential that the divergence of
individual interests causes. Unfortunately, many, if not most, of these losses

are too subtle to be captured in any economic data. Nonetheless, they are of immense importance.

Today, economists are becoming interested in one aspect of such matters, which they label "asymmetrical information." In 1970, George Akerlof, an economist from the University of California at Berkeley, published a landmark article bearing the strange title "The Market for 'Lemons': Asymmetrical Information and Market Behavior." Akerlof, who had spent time in India, was struck by the difficulty of doing business in the Third World, yet his example of lemons indicated that the same problems persisted in the developed countries.

Akerlof's lemons were used cars of dubious quality. The "asymmetrical information" in the title, then, refers to the fact that the seller of the used car knows for sure whether it is a lemon, but the buyer lacks this information.

The buyer is left to infer the motives of the seller. Most buyers wonder why someone should want to sell a good quality used car. Shouldn't the owner want to keep it so long as it has no problems? The mere fact that the car is on the market suggests that it is a lemon. By this logic, anyone with enough money to avoid buying a used car would do so, reducing the demand for used cars and expanding the demand for new ones.

The missing element in the market for lemons is trust. After all, mistrust is almost synonymous with the popular reputation of used car salesmen. When Richard Nixon ran for governor of California, one of the most effective campaign advertisements used against him was a picture of the candidate with the question, "Would you buy a used car from this man?"

Since Akerlof's article first appeared, economists have published a virtual torrent of works on the subject of asymmetric information. The majority of these have shown, in the spirit of Akerlof's article, how doubts about the intentions or aptitudes of buyers, borrowers, or employees make markets malfunction.

Of course, Akerlof was not the first to recognize the economic importance of trust. More than three centuries ago, at the dawn of our market society, the seventeenth-century philosopher, John Locke, declared trust to be "the bond of society," the *"vinculum societatis."* Although he may have exaggerated a bit, he was no doubt on the mark when he maintained that self interest cannot be the basis of "the law of nature" (Locke 1663, p. 213).

John Locke's contemporary, Thomas Hobbes, penetrated the essence of the importance of trust, observing:

> [The] force of Words . . . [is] too weak to hold men to the performance of their Covenants; there are in man's nature, but two imaginable helps to

strengthen it. And those are either a Feare of the consequence of breaking their word; or a Glory, or Pride in appearing not to need to breake it. (Hobbes 1651, p. 200)

Hobbes recognized that within the emergent capitalist society of his time, the benefits of Glory or Pride did not match the opportunity to make a quick buck. He concluded that "there must be some coercive Power, to compel men equally to the performance of their Covenants, by terrour of some punishment" (p. 202).

This necessary terror requires a strong state. Without such a power, "there is . . . a perpetuall warre of every man against his neighbour; And therefore everything is his that getteth it and keepeth it by force, which is neither *Propriety* nor *Community,* but *Uncertainty*" (p. 296).

No state, however strong it may be, can coerce people to be trustworthy. Even so, Hobbes went to the heart of a serious problem in a market society.

In recent years, the sharp insight of Hobbes has given way to a shallow cynicism about the nature of trust. The fashionable new amalgam of economics and legal theory, known as the Law and Economics School theory, advocates a doctrine of "efficient breach." According to this theory, people should subject legal agreements to a market test. A party to a contract might be well advised to break it, if the penalties are light enough to make the breach cost-effective to do so (Kuttner 1997, p. 64–65). This perspective is particularly interesting in light of the earlier discussion about deterrence and the lax enforcement of corporate malfeasance.

While the theory of efficient breach makes economic sense for an isolated individual, it is suicidal for society as a whole. Trust cannot flourish in such an environment.

Unfortunately, we may now reside in such an environment. Robert Putnam provides a suggestive statistic about the breakdown in trust. He reports that the proportion of Americans saying that most people can be trusted fell by more than a third between 1960, when 58 percent chose that alternative, and 1993, when only 37 percent did (Putnam 1995, p. 72). Alas, in our own world of hypercapitalism, Thomas Hobbes sounds far more realistic than Alfred Marshall's quaint depiction, to which we will now turn.

The Economics of Information and Trust

Alfred Marshall was probably the most influential economist of the early twentieth century. In Marshall's day, Hobbes's skepticism about the possibility of voluntary compliance with contracts might have seemed unduly cynical. Marshall himself believed that developing a reputation for trust-worthiness made good business sense. He wrote:

> A producer, a wholesale dealer, or a shopkeeper who has built up a strong connection among purchasers of his goods, has a valuable property. . . . [He] expects to sell easily to them because they know and trust him and he does not sell at too low prices in order to call attention to his business, as he often does in a market where he is little known. (Marshall 1923, p. 102)

Marshall was at least partially correct for his time. A solid reputation would have considerable value for Marshall's shopkeeper, unlike an itinerant ped-dler, who might never see his customers again. The shopkeeper's customers were probably members of his own community. He and his family would have frequent encounters with them. To earn their enmity would under-mine the possibility of an ongoing business relationship, as well as dimin-ish the quality of the shopkeeper's life. The more dealers nurture their reputations, the more effective information buyers will have about the markets that they face.

In a world with undreamt-of mobility, the consequences for betraying a trust have greatly diminished. Perhaps such changes explain why, since Marshall's day, economists have, with few exceptions, largely ignored the question of trust. Albert Hirschman did point out that trust, like informa-tion, increases with use (Hirschman 1984, p. 93).

Kenneth Arrow put the economics of trust in a somewhat broader per-spective, observing:

> In the absence of trust it would be very costly to arrange for alternative sanctions and guarantees, and many opportunities for mutually beneficial cooperation would have to be foregone. Banfield has argued that lack of trust is indeed one of the causes of economic underdevelopment. (Arrow 1971; referring to Banfield 1958)

Similarly, Samuel Bowles and Herbert Gintis noted:

> In the absence of trust it would be very costly to arrange for alternative sanctions and guarantees, and many opportunities for mutually beneficial cooperation would have to be foregone . . . norms of social behavior, in-cluding ethical and moral codes (may actually be) reactions of society to compensate for market failures. (Bowles and Gintis 1995)

Charles Sabel, influenced by his study of market relations in small Italian towns, which may have had more in common with Marshall's England than the modern day United States, proposed a rather different understanding of trust. For him: "Trust . . . is like a constitutional, democratic compact which requires of the parties only that they agree to resolve disputes in ways that do not violate their autonomy, and roots this agreement in the citizen's recognition of the connection between the assertion of one's own autonomy and respect for that of others (Sabel 1993, p. 1143).

Finally, Oliver Williamson, whom we shall discuss later, describes the expanding scope of modern corporations as an effort to cope with "the full set of *ex ante* and *ex post* efforts to lie, cheat, steal, mislead, disguise, obfuscate, feign, distort and confuse" (Williamson 1985, p. 51 fn).

While these authors deserve credit for raising the question of trust, we might fault them for failing to recognize that trust represents a deeper layer of the economy. Trust has profound consequences that economists ignore, probably because they are not at all amenable to the typical tools of economic analysis.

On the most superficial level, while trust cannot completely replace the police departments, courts, and prisons, it can minimize the need for such institutions. Trust can make work more pleasant, amplifying people's productivity. Finally, a trusting environment can give people the confidence to explore new possibilities, allowing the economy to come closer to its full potential. I believe that Tawney had such possibilities in mind when he argued for a more egalitarian society.

Despite the fact that a few of the best academic authors realize that the development of trust is a necessary component of a well-functioning economy, trust is far less likely in a rapidly changing society, where culture and tradition count for little. Today, we mostly deal with faceless corporations, whose ownership can change at a moment's notice.

Besides, financial pressures require that immediate profits be maximized, even if the necessary actions might undermine future profits. When economic conditions are so fluid and people have the chance to make considerable money within a short period of time, they will be more likely to "sell" their reputation for the chance to make a quick killing. For example, some corporations declare strategic bankruptcy in order to caste aside their pension obligations to their employees (Orr 1998).

CHAPTER 5

Conflict in the Production Process

Contested Terrain

When we look at society, we see that people who give orders rank above those who take orders. To give another the right to give us orders is to descend in rank voluntarily. Few of us would readily do so.

A certain degree of resentment usually accompanies the necessity of taking orders. The resulting tensions between labor and capital give rise to a major source of waste. Some might argue that we should not have regard for the losses due to conflicts between labor and capital because such tensions are inherent in the human psyche rather than a reflection of the nature of our economy.

In reality, the degree to which authoritarianism is inherent in the employment relationship is not absolute. For example, today, with job insecurity probably more pervasive in the United States than it has been at any time since the depths of the Great Depression, labor has become much more pliant than it was in the 1960s and early 1970s.

I would like to explore this dimension of waste now. Perhaps nobody delved more deeply into the conflicts created by the employment relationship than Karl Marx. Yes, I know that Marx is supposed to be passe today. Bear with me for a few moments, anyway, while I refer back to Marx. I will indicate the pertinence of his analysis in due course.

For Marx: "all means for the development of production . . . distort the worker into an appendage of a machine, they destroy the actual content of his labour by turning it into a torment; they alienate from him the intellectual potentialities of the labour process; . . . they deform the conditions under which he works, subject him during the labour process to a despotism more hateful for its meanness" (Marx 1977, p. 799).

Consequently, wrote Marx: "the worker actually treats the social character of his work, its combination with the work of others for a common goal, as a power that is alien to him; the conditions in which this combination is realized are for him the property of another" (Marx 1981; 3, pp. 178–9).

Now, consider how Marx's insights hold up in a modern production facility.

Conflict on the Assembly Line

People do not like taking orders. When employment conditions become more favorable to labor, workers become emboldened. At such times—especially when workplace authorities do not treat workers with respect, workers sometimes confront management in a more direct form, often going to great lengths to exercise some control over the labor process. Workers may resist authority, even when they have no expectation of wringing any concessions from management. The idea of exercising control, even when that control is nothing more than the disruption of the labor process, can be a source of delight to workers who have little other discretion over their job.

The testimony of John Lippert, who had been a worker in General Motors' Fleetwood plant, offers some insight into worker's efforts to assert some control over their lives while on the job. He described his fellow workers' vigorous efforts to slow down the assembly line (Lippert 1978). His obvious pride in hurrying up his work on a few cars, in order to have 15 or 20 seconds to himself suggests the intensity of his desire to carve out even small bits of autonomy on the assembly line (Lippert 1978, p. 58).

Bill Watson, who spent one year in a Detroit automobile factory, offers an even more dramatic example of the lengths to which workers go to assert their independence (Watson 1971). He describes how workers revolted against the production of a poorly designed six-cylinder model car. After management rejected the workers' suggestions for improvements in the production and design of the product, they initiated a "counter-plan," beginning with acts of deliberately misassembling or omitting parts on a larger than normal scale. Later, workers in inspection made alliances with workers in several assembly areas to ensure a high rate of defective motors. Eventually, even more complicated measures were taken.

In the process, workers and foremen argued over particular motors. Tension escalated. Instead of just building defective motors, workers went ahead and installed many of them, thereby requiring that management would have to go to the trouble and expense of removing them later. The conflict only ended during a layoff after management suddenly moved the entire six-cylinder assembly and inspection operation to another end of the plant, presumably at great cost (Watson 1971, pp. 76–77).

In another instance, the company, intending to save money by shutting down their foundry early, attempted to build the engines using parts that already had been rejected during the year. Workers in the motor-test area lodged the first protest, but management hounded inspectors to accept defective motors. After motor-test men communicated their grievances to other workers, they began to collaborate in intentional sabotage. Inspectors agreed to reject three of every four motors. Stacks of motors piled up at an accelerating pace until the entire plant shut down, losing more than 10 hours of production time to deal with the problem. Management summoned inspectors to the head supervisor's office. The inspectors slyly protested that they were only acting in the interest of management.

Watson's third example is the most telling of all. During a model changeover period, management had scheduled an inventory buildup, which was to require six weeks. Management intended to keep 50 people at work on the job. These workers would have earned 90 percent of their pay if they had been laid off. Workers reacted to the opportunity, attempting to finish the inventory in three or four days instead of the six weeks. They trained each other in particular skills, circumventing the established ranking and job classification system to slice through the required time.

Management responded harshly, forcing workers to halt, claiming that the legitimate channels of authority, training, and communication had been violated. If workers had been given the opportunity to organize their own work, Watson claims that they could have completed the task in one-tenth the required time. Management, however, was determined to stop workers from organizing their own work, even when it would have been finished quicker and management would have saved money (Watson 1971, p. 80). So much for the idea that market forces lead people to choose rationally!

Watson also described how workers engaged in hose fights and even organized contests to explode rods from engines in the workplace. These incidents illustrate the enormous costs associated with a conflictive system of labor relations. One might argue that the managers that Watson described were unusually short-sighted, but I suspect that they were merely less subtle.

To admit that workers have something to contribute besides merely carrying out the demands of management undermines, in part, the ultimate rationale for management's authority. As a result, managers typically resist all encroachments on their authority.

Watson communicates a sense of intense joy and exhilaration that workers felt from having the opportunity to organize their own activity. This motivation would still remain, even if capitalists were not as shortsighted as those that Watson described. He applauded industrial sabotage as "the forcing of more free time into existence" (Watson 1971, p. 80). He explained:

> The seizing of quantities of time for getting together with friends and the amusement of activities ranging from card games to reading or walking around the plant to see what other areas are doing is an important achievement for the laborers. Not only does it demonstrate the feeling that much of the time should be organized by the workers themselves, but it also demonstrates an existing animosity. . . .
>
> While this organization is a reaction to the need for common action in getting the work done, relationships like these also function to carry out sabotage, to make collections, or even to organize games and contests which serve to turn the working day into an enjoyable event. (Watson 1971, pp. 80–81)

Even in cases where contemporary workers might be the nominal owners of the firms, they still resist management's authority, especially when they regard it as unjust. Consider the case of Rath Meatpacking, a firm that the workers had bought out. Although the workers were the nominal owners, the structure of authority remained unchanged. According to a *Wall Street Journal* article:

> Frustration at Rath has progressed to the point of sabotage. One employee expressed his anger by deliberately allowing the hair-picking machine that he runs to clog up so as to force his side of the hog-kill floor to shut down for the day. "My foreman didn't have any idea how to fix it," he says. "This is one way to show him who's boss." (Minsky 1981; see also Brecher 1988)

I realize that the examples that I have just used are somewhat dated, coming from a time when job security was much higher. Again, I would ask that you bear with me a bit longer while I briefly return to Marx again. Soon thereafter, I will attempt to address more subtle aspects of the subject of the costs of conflict between labor and capital.

Marx and the Cost of Opposition

Some readers might be tempted to interpret the materials from the previous section as suggesting that stronger authority relationships would improve efficiency. I would like to offer a different reading. While hose fights and some of the other antics of the 1970s might seem unlikely today, Marx insisted that people on the outside have difficulty in getting a clear picture of what goes on in what he called "the hidden abode of production" (Marx 1977, p. 279).

However, during times when labor feels confident enough to challenge capital openly, we can see evidence of conflict in the workplace that otherwise would remain obscured. The opposition between labor and capital creates unavoidable costs. In Marx's words:

> The exploitation of labour costs labour. Insofar as the labour performed by the industrial capitalist is rendered necessary only because of the contradiction between capital and labour, it enters into the cost. . . . in the same way as costs caused by the slave overseer and his whip are included in the production costs of the slave owner. (Marx 1963–1971, Pt 3, p. 355)

Marx claimed no particular originality in making this point. He quoted a widely read student of the early evolution of the manufacturing system, Andrew Ure, who wrote, "By the infirmity of human nature, it happens that the more skilful the workman, the more self-willed and intractable he is apt to become" (Ure 1835, p. 20; cited in Marx 1977, p. 490). As a result, capitalists are forced to use a significant amount of resources to attempt to subdue workers' resistance. Marx observed:

> The work of directing, superintending and adjusting becomes one of the functions of capital, from the moment that the labour under capital's control becomes co-operative. As a specific function of capital, the directing function acquires its own special characteristics. . . .
>
> As the number of operating workers increases, so too does their resistance to the domination of capital, and necessarily, the pressure put on by capital to overcome this resistance. The control exercised by the capitalist is not only a special function arising from the nature of the social labour process, and peculiar to that process, but it is at the same time a function of the exploitation of a social labour process, and is consequently conditioned by the unavoidable antagonism between the exploiter and the raw material of his exploitation. (Marx 1977, p. 449)

The more effective employers are in wielding power in the workplace, the more workers will turn to other means to challenge capital. For example, in industries where the employers' power was most absolute, such as lumber and mining, workers tended to turn to militant organizations, such as the International Workers of the World. In addition, when employers' power in the workplace is stronger, workers are more likely to turn to political struggles. In this context, we would do well to recall the great importance that Marx placed on labor's ongoing struggle to limit the extent of the working day, one of the most burning issues of the time (see Marx 1977, Chapter 10).

Marx believed that such actions would be more effective in the long run than strikes or other direct actions against individual employers or even industry associations, because political efforts would make workers see themselves as a class. Marx believed that only after workers felt more loyalty to their class than to their trade, would they become a more effective unit of resistance.

Marx realized more than a century ago that, even if workers' resistance does not take forms as obvious as hose fights, capitalists will still be forced to devote considerable resources to monitoring workers' performance. As Marx wrote that:

> one part of the labour of superintendence merely arises from the antagonistic contradiction between capital and labour, and from the antagonistic character of capitalist production. (Marx 1963–1971, Pt 3, p. 505)

Thus, Marx noted:

> The less he is attracted by the nature of the work and the way in which it has to be accomplished, and the less, therefore, he enjoys it as the free play of his own physical and mental powers, the closer his attention is forced to be. (Marx 1977, p. 284)

Or again, Marx observed that:

> this work of supervision necessarily arises in all modes of production that are based on opposition between the worker as direct producer and the proprietor of the means of production. The greater this opposition, the greater the role that this work of supervision plays. . . .
> This work of management and supervision, in so far as it is not simply a function arising from the nature of all combined social labour, . . . arises

rather from the opposition between the owner of the means of production and the owner of mere labour-power. (Marx 1981; 3, pp. 507 and 509)

Marx was not alone in his concern for such matters. John Stuart Mill, the most eminent British economist of the mid-nineteenth century, came to very similar conclusions.

John Stuart Mill and the Cost of Opposition

John Stuart Mill was a most unusual economist. The son of a well-placed writer on economics, Mill grew up surrounded by some of the most famous intellectuals of his day, including the most important economists of Britain. Mill was sympathetic to the plight of the workers, and even said that he wanted to see a socialistic type of society, but in typical conservative fashion, he held the workers responsible for their sorry state of affairs.

> The deficiency of practical good sense, which renders the majority of the labouring class such bad calculators—which makes, for instance, their domestic economy so improvident, lax, and irregular—must disqualify them for any but a low grade of intelligent labour, and render their industry far less productive than with equal energy it otherwise might be. (Mill 1848, I.vii.5, p. 107)

A couple of pages later, he wrote:

> As soon as any idea of equality enters the mind of an uneducated English working man, his head is turned by it. When he ceases to be servile, he becomes insolent. . . . The moral qualities of the labourers are fully as important to the efficiency and worth of their labour, as the intellectual. Independently of the effects of intemperance upon their bodily and mental faculties, and of flighty, unsteady habits upon the energy and continuity of their work (points so easily understood as not to require being insisted upon), it is well worthy of meditation, how much of the aggregate effect of their labour depends on their trustworthiness. (Mill 1848, I.vii.5, p. 109)

Then, in his very next sentence, Mill began to touch upon many of the themes of this book, observing:

> All the labour now expended in watching that they fulfil their engagement, or in verifying that they have fulfilled it, is so much withdrawn from the real business of production, to be devoted to a subsidiary function rendered

needful not by the necessity of things, but by the dishonesty of men. Nor are the greatest outward precautions more than very imperfectly efficacious, where, as is now almost invariably the case with hired labourers, the slightest relaxation of vigilance is an opportunity eagerly seized for eluding performance of their contract. The advantage to mankind of being able to trust one another, penetrates into every crevice and cranny of human life: the economical is perhaps the smallest part of it, yet even this is incalculable. To consider only the most obvious part of the waste of wealth occasioned to society by human improbity; there is in all rich communities a predatory population, who live by pillaging or overreaching other people; their numbers cannot be authentically ascertained, but on the lowest estimate, in a country like England, it is very large. The support of these persons is a direct burthen on the national industry. The police, and the whole apparatus of punishment, and of criminal and partly of civil justice, are a second burthen rendered necessity by the first. The exorbitantly-paid profession of lawyers, so far as their work is not created by defects in the law, of their own contriving, are required and supported principally by the dishonesty of mankind. As the standard of integrity in a community rises higher, all these expenses become less. But this positive saving would be far outweighed by the immense increase in the produce of all kinds of labour, and saving of time and expenditure, which would be obtained if the labourers honestly performed what they undertake; and by the increased spirit, the feeling of power and confidence, with which works of all sorts would be planned and carried on by those who felt that all whose aid was required would do their part faithfully according to their contracts. Conjoint action is possible just in proportion as human beings can rely on each other. There are countries in Europe, of first-rate industrial capabilities, where the most serious impediment to conducting business concerns on a large scale, is the rarity of persons who are supposed fit to be trusted with the receipt and expenditure of large sums of money. There are nations whose commodities are looked shyly upon by merchants, because they cannot depend on finding the quality of the article conformable to that of the sample. Such short-sighted frauds are far from unexampled in English exports. (Mill 1848, I.vii.5, p. 109)

We find reference to something akin to the comparison of efficiencies of various social infrastructures that Jones and Hall: the waste involved in monitoring workers and the waste of the legal system, including the cost of the system of criminal justice. However, intermingled with these thoughts comes the claim that a good deal of the waste would not occur if "the labourers honestly performed what they undertake." Although this long paragraph also found fault with the dishonesty of other elements of society, Mill, despite his self-proclaimed socialist views, concentrated his ire on the deficiencies of the working class.

The Short End of the Stick and the Short-Handled Hoe

In asserting that the conflicting goals of workers and supervisors create enormous waste in the workplace, I have used relatively flagrant examples of worker resistance from the automobile industry during the 1970s. I will now turn to two more examples to provide a further glimpse of the tensions within the workplace.

Let me first consider the example of farm workers in California who worked under employers who enjoyed far more dominance over their employees than the automobile companies I discussed earlier. Despite the absence of workers' rights on farms, employers still had to find ways to ensure that they could get as much work as possible from their laborers. This need for supervision distorted the economy in ways that might not be obvious at first sight. For example, until widespread public protests forced California to outlaw short-handled hoes, farmers in the state required their farm workers to use these unwieldy implements that forced people to stoop down while working. The short-handled hoe is no more efficient for hoeing than the more familiar hoe—at least in a technical sense.

The only advantage of the short-handle is that it facilitates monitoring. Because of the inadequate length of the handle, this tool forces the worker to stoop in an unnatural position that places a great strain on the back. This stooping works to the advantage of the employer. The foreman can easily see when workers are relaxed, since they naturally stand erect to relieve the pressure on their backs whenever possible. As would be expected, the continual bending frequently caused serious back injuries. Nonetheless, farming interests fought vigorously to protect their continued use of the short-handled hoe, insisting that the hoe was perfectly adapted to the physiological characteristics of the immigrant workers (see Perelman 1977).

Society at large never had an inkling of the human cost of the short-handled hoe until the activities of the United Farm Workers Union brought the farm workers' plight out of the "hidden abode" and into the public view. At the time, people casually referred to farm work as stoop labor. What could be the harm in using such a tool? In reality, the bent backs of the farm workers reflected the drive to extract as much value as possible from the employment of people who had precious few other options.

The use of ninety-nine-cent pricing illustrates a more subtle means by which the need for supervision distorts the economy. The invention of the cash register helped store owners prevent employee theft, since the register kept a record of each transaction that the employee rang up. The registers were not fool-proof, however, since employees still had the option of not ringing up the sale and then pocketing the money for themselves.

With this problem in mind, employers turned to ninety-nine-cent pricing, which became common soon after the introduction of the cash register. With ninety-nine-cent pricing, customers would be less likely to pay the exact price. The clerk, in turn, would then be more likely to open the cash register to get a coin, which could only be done by ringing up the sale (Huston and Kamdar 1996, pp. 137–38).

Over and above the cost of installing a cash register, the need for supervision distorts the price structure (albeit by a small amount), increases the work done by clerks (in terms of handing out change), and causes extra bookkeeping costs for consumers, who balance their checkbooks manually (assuming that the rounded numbers would be easier to tally). Of course, the penny in change works to the advantage of consumers. In effect, the merchant paid the customer a penny to help monitor workers.

While the costs associated with ninety-nine-cent pricing might not be large, they are also far from obvious, suggesting that many other aspects of the cost of supervision go unnoticed.

In the examples of both the short-handled hoe and ninety-nine-cent pricing, we encounter employers taking measures to oppose workers' resistance to following the dictates of the employer. In the one case, the workers wish to enjoy a few moments of rest on the job; in the other case the workers take some of the employers' money. In either case, the employer regards the employee's act as a form of theft, taking something that is rightfully his, whether it be money or working time.

In the case of the farm workers, the burden literally falls on the backs of the laborers. In the case of the clerks in the store, they have to make change without any technical reason to do so. In the case of the farm workers, the short-handled hoe is no more expensive than the normal hoe. In the case of the clerks, the owners of the stores incurred the extra expense of the cash register in order to monitor the workers.

The short-handled hoe fell into oblivion, while the cash register company pioneered the revolution in office machinery. Not entirely coincidentally, the founder of IBM began with the National Cash Register Corporation, and ended by leading an industry that is making an enormous number of clerks obsolete while facilitating ever more extensive methods of monitoring workers as well as society at large.

Subtle Resistance within the Labor Process

Let us continue with the theme of the subtle distortions of the labor process. Because workers resent their condition for all the reasons given in the previous section, they subvert the labor process in ways other than obvious resistance to management's authority. Workers' covert resistance covers a broad continuum ranging from outright sabotage to feigning incapacity, such as the "Sambo" behavior that Stanley Elkins attributed to slaves (Elkins 1968, pp. 131 ff).

The Sambo phenomenon rests on a single paradox: if the underling is supposed to be inferior and stands in need of direction from superiors, then how can he or she be expected to perform in a responsible fashion? Eugene Genovese's *Roll, Jordon, Roll* explored the remarkably subtle means by which slaves were able to exert their considerable influence within a society that refused to recognize them as full human beings (Genovese 1976). This influence extended beyond the fields into every walk of life.

Jaroslav Hasek's novel, *The Good Soldier Sveijk,* the charming story of a bohemian soldier caught up in the turmoil of the World War I offers a particularly delightful example of the Sambo phenomenon (Hasek 1974). Most of the humor in the book arises from Sveijk's practice of subverting the relationship with his superiors. He would seize upon some ambiguity or hyperbole in the orders that his superiors give him, and then take that aspect of his orders literally in such a way that he was able to make the authority relation ridiculous. By this means, Sveijk was able to do virtually whatever he wanted to do. When challenged to explain his behavior, Sveijk unflinchingly boasted of being a lunatic, much to the consternation of his superiors and to the delight of generations of readers. The same defect that drove Sveijk's officers to distraction continues to plague the employer/employee relationship.

Sveijk-like behavior is not restricted to the world of fiction. Workers who feel resentful frequently take advantage whenever possible of any ambiguity within the workplace. One strategy is to learn to perform their tasks with a minimum of exertion (Marx 1977, p. 458). Even on the slave plantations of the southern United States, where authority over the workers was all but absolute, slaveholders had to use heavy, inefficient farm implements and mules instead of more efficient, lighter, horse-drawn implements to minimize slaves' sabotage (Marx 1977, p. 304). On a hot, muggy day while the slavedriver glances away, a slave might be tempted to "stupidly" damage a horse or a piece of equipment in order to take a break from hard labor.

Stanley Mathewson reported a classic description of an automobile worker's finding a loophole in a job description worthy of the good soldier Sveijk:

> A Mexican in a large automobile factory was given the final tightening to the nuts on automobile-engine cylinder heads. There are a dozen or more nuts around this part. The engines passed the Mexican rapidly on a conveyer. His instructions were to test all the nuts and if he found one or two loose to tighten them, but if three or more were loose he was not expected to have time to tighten them.

[A supervisor who was puzzled that so many defective engines were passing along the line] discovered that the Mexican was unscrewing a third nut whenever he found two already loose. (Mathewson 1939, p. 125)

After all, loosening one nut required less effort than tightening two. A famous railroad manager and disciple of Frederick Taylor, Harrington Emerson, related a similar incident: "A railroad track foreman and gang were recently seen burying under some ashes and dirt a thirty-foot steel rail. It was less trouble to bury it than to pick it up and place it where it could be saved" (Emerson 1912, p. 67; cited in Palmer 1975, p. 37).

Workers creatively find other methods of subverting management's authority. For example, they can ostensibly obey management by following practices to the letter, which generally causes a slowdown. Harley Shaiken uses the example of a machine shop:

> A familiar sight in most shops is an engineer walking in with a stack of blueprints to ask the worker if a particular job is feasible. The machinist carefully studies the prints, looks at the engineer, and says, "Well, it can be tried like this but it will never work." Grabbing a pencil, the machinist marks up the print and, in effect, redesigns the job based on years of experience. . . .
> [In one shop, when] management initiated a campaign to strictly enforce lunch periods and wash-up time, the judgement of some machinists began to fade. About this time a foreman dashed up to the shop with a "hot" job. . . . Anxious to get the job done quickly, the foreman insisted that the machinist run the lathe at a high speed and plunge the drill through the part. Under normal circumstances the machinist would have tried to talk the foreman out of this approach but now he was only too happy to oblige what were, after all direct orders. The part not only turned out to be scrap,

but part of the lathe turned blue from the friction generated by the high speed. The disciplinary campaign was short-lived. (Shaiken 1985, pp. 19–20)

In one case, pilots at Eastern Airlines reversed the classic strategy of a slow-down. They pressured their employer by flying at higher speeds, which burnt more fuel. This extra cost undermined corporate profitability (Anon 1988).

Good Help Is Hard To Find

Sveijk's ploy represents one dimension of what modern economics calls "the principal/agent problem." To ensure that a subordinate carries out orders, the employer must first convey orders in a clear and unambiguous manner. Sveijk's exploits suggest that the effective framing of orders is no easy matter.

Even if managers succeed in giving clear and unambiguous orders, and the workers understand what is expected of them, management must still find a means to make workers carry out their tasks in a satisfactory manner. In the face of the complexity of the labor process, employers cannot be sure that workers are acting in the best interest of the firm, even when they are trying to observe them carefully. Moreover, attempts to collect information on the workers' performance are costly. The ability to collect information is made even more difficult because workers often attempt to distort the flow of information within the firm to gain a strategic advantage (see Annable 1988).

Given workers' resistance, how can the direct authority of employers ever succeed in directing the labor process in all its complexity, especially since capitalists and their agents are so convinced of their overwhelming superiority? For the most part, such complications have eluded mainstream economists who prefer to overlook the intricacies of the labor process.

Recently, some economists have begun to take an interest in the labor process. They have discovered that the labor process is far too complex for any employer to give complete and precise instructions in advance. No manager could presume to know all the contingencies that will arise in the course of the working day (see Hart 1988, p. 469).

Moreover, when capitalists do exert themselves in an attempt to monitor the production process more intensively, workers consider those efforts intrusive. As a result, any benefits that management might enjoy from superior information may be dissipated in increased workers' resistance. For example, the early efforts of Ford Motor Company's Sociological Department to develop intricate controls over workers' lives during the 1930s, in an attempt that some Europeans called Fordism. Although they believed that this sort of domination represented a harbinger of the future, it was not duplicated elsewhere, in part because of the resistance that it provoked (Sward 1972, pp. 58–60; see also Aglietta 1979, pp. 116–30). To make matters worse, even if unlimited supervision were both possible and affordable, it would be still unlikely to succeed without some goodwill on the part of the workers.

For example, F. M. Scherer, dean of those economists who specialize in studying economic organization, has amassed evidence to support his contention that the higher wages paid in large plants does not reflect the monopoly power that such employers enjoy, as most economists believe, but rather such wage premia are a payment for the higher alienation that workers experience in such environments (Scherer 1976). Belcher's text, *Wage and Salary Administration,* made a similar point:

> Labor is not a simple commodity . . . that can be bought according to specifications. . . . Labor is not only perishable; since it varies with the ability of a person to do work, it may be a different commodity from day to day or hour to hour, and is affected by all manner of things. Management recognizes that part of the [worker's] price may go toward promoting a feeling of loyalty to the organization and enthusiasm for work, thereby stimulating the worker to be a more efficient worker. (Belcher 1962, p. 4; cited in Annable 1988, p. 12)

In fact, Henry Ford's famous Five Dollar Wage plan was not, as he falsely claimed, an enlightened policy to allow workers to buy back the cars that they produced; instead, it was a response to the restiveness of his workers in the face of the extraordinary demands that Ford placed on them (see Meyer 1981, p. 1; and Raff and Summers 1987). Ford himself hinted at his real motivation, writing:

> There was . . . no charity involved. . . . We wanted to pay these wages so that business would be on a lasting foundation. We were building for the future. A low wage business is always insecure. The payment of $5 a day for an 8 hour day was one of the finest cost cutting moves we ever made. (Ford 1922, pp. 126, 127, and 147; cited by Raff and Summers 1987, p. S60).

Ford's logic is relatively straightforward. When workers feel exploited, they take measures to try to get even. For example, Sam Bowles and Rick Edwards reported on a 1983 study by the United States Department of Justice that found that more than two-thirds of workers in the United States engage in counterproductive behavior on the job, including excessively long lunches and breaks, slow, sloppy workmanship, and sick-leave abuse as well as the use of alcohol and drugs on the job. One-third of a sample of 9,175 randomly selected retail, manufacturing, and hospital workers admitted stealing from their employers. In-depth interviews with a smaller sample revealed that the workers were responding to a feeling of being exploited rather than dire economic necessity (Bowles and Edwards 1985, p. 179).

A Battle of Wills or Using Workers' Skills

Supervisors supposedly get to supervise because they "know" more than the lower-ranking employee. In many cases, however, management just does not understand the production process enough to make it run very well.

Few employees are inclined to do much to correct the situation. Why should they? After all, when workers speak up, their input is often unwelcome. Supervisors often respond to suggestions from employees by saying something such as, "We pay you to work, not to think." If supervisors do listen, employees are unlikely to benefit directly; instead, they might risk making themselves or their fellow workers superfluous.

Nonetheless, workers typically have considerable skills. They can apply them to making the workplace more efficient or they can use their skills to subvert the production process. For example, Watson's previously discussed description of counterplanning in the automobile industry implies that even workers, who are contemptuously called "semi-skilled," were more aware than higher management of the deficiencies of the motor that they were manufacturing. Such workers are in possession of valuable information. Doeringer and Piore wrote:

> Almost every job involves some specific skills. Even the simplest custodial tasks are facilitated by familiarity with the physical environment specific to the workplace in which they are being performed. The apparently routine operation of standard machines can be importantly aided by familiarity with the particular piece of operating equipment. . . . In some cases workers are able to anticipate trouble and diagnose its source by subtle changes in the sound or smell of the equipment. Moreover, performance in some production or managerial jobs involves a team element, and a critical skill in the ability to operate effectively with the given members of the team. This ability is dependent upon the interaction skills of the personalities of the members, and the individual's work "skills" are specific in the sense that skills necessary to work on one team are never quite the same as those required on another. (Doeringer and Piore 1971, pp. 15–16)

Consider Ernesto Galarza's analysis of the skills required for agricultural field work, a class of labor thought to be among the most unskilled known to modern society:

> [T]hose who persisted in stereotyping all harvest hands as "unskilled" failed to acknowledge many skills that proper handling of tomatoes or cantaloupes or peaches required. These skills, practiced with the economy of motion and effort that only experience can bring and refined into a kind of wisdom of work, the domestics possessed. . . .
>
> Field labor was a blur in which the details of field harvesting and the skills it required went unrecognized. To pick a ripe honeydew requires a trained eye for the bloom of tinted cream, a sensitive touch for the waxy feeling of the rind, and a discriminating nose for the faint aroma of ripeness. In the asparagus fields, the expertness of the Filipino cutters was obvious to all but those who hired them. (Galarza 1977, pp. 29 and 366)

Doeringer and Piore pointed out that the specialized job knowledge provides labor with a strategic advantage in the workplace, especially in a dynamic technological environment:

> When the job is specific, the workman tends to have a monopoly over a portion of the knowledge required to maintain and operate the technology. The importance of experienced workmen in the process of on-the-job training stems in part from this. But the monopoly also gives independent power to disrupt the production process. Given the fact that the technology is unwritten, and that part of the specificity derives from improvements which the work force itself introduces, workers are in a position to perfect their monopoly over the knowledge of the technology should there be an incentive to do so. (Doeringer and Piore 1971, p. 84)

A truly efficient economy would be able to tap into the unappreciated skills of the workforce. Instead, as we have seen, workers lacking a creative outlet for their skills sometimes use their skills to subvert the production process. Mathewson devoted an entire book to documenting the extent to which workers are able to restrict their output, especially when they feel personally aggrieved (Mathewson 1939).

Mathewson noted that automobile workers' earnings often rose appreciably just before the end of a production run for a particular model. The reason for this change in behavior is understandable. Firms would not have enough time to change the rates on the old model. Moreover, the workers' more rapid pace would not be relevant to the setting of rates on the new model (Mathewson 1939, pp. 61–62; also cited in Clark 1984). At the turn of the century, a report of the United States Commissioner of Labor found that output restriction was typically between 13 and 50 percent even though many of the industries were not formally organized (United States Commissioner of Labor 1904; cited in Clark 1984, p. 1074).

Norsworthy and Zabala's econometric exercise provided further evidence of labor's ability to restrict its efforts. In their study, they developed an index of worker attitude based on plant level data on grievances, quits, and unauthorized strike activity. Their index is a statistically significant variable in explaining movements in automobile workers' productivity (Norsworthy and Zabala 1985). Similarly, Michele Naples has used quit rates and industrial accidents as proxies for workers' resistance to their conditions. These two variables may go a long way toward explaining the decline in United States manufacturing productivity that occurred during the 1970s (Naples 1988).

These regressions, as well as Watson's description of the activities of his fellow automobile workers, call into question the notion of management as an effective organizer of the labor process. While many economists would be more than willing to grant that "unruly" labor can disrupt the production process, the economics literature does not even hint that workers might have any potential to bring their own creativity into the production process. Supposedly only management has the intelligence to regulate production.

In conclusion, instead of being a one-sided affair, the labor process is a complex contest in which both labor and capital use all the resources at their disposal to gain an advantage. In part, both labor and capital apply ingenuity to make the conditions of production more to their liking. True, capitalists' power is more obvious. In many cases, they can fire workers or reduce their wage at will, but that form of power is only one aspect of the two-sided labor process.

In the course of these epic struggles between labor and capital, a considerable, but immeasurable amount of the resources directly devoted to the production of goods and services waste away. These losses can take the form of harm to the workers themselves, as in the case of the farm workers using the short-handled hoe, resources consumed in monitoring workers, lost production, or even destruction of output. The greatest loss, I am convinced, is the failure to take advantage of and to nurture the skills, and even more so the creativity of the unappreciated people who toil to supply us with the goods and services that make up our standard of living.

Strategies of Supervision

Monitoring of workers consumes an unbelievable amount of time and re-sources. Workers in the United States may be the most oversupervised in the world. In his wonderful book, *Fat and Mean,* David Gordon reported:

> Depending on the definition, between 15 and 20 percent of private non-farm employees in the United States work as managers and supervisors. In 1994 we spent $1.3 trillion on the salaries and benefits of nonproduction and supervisory workers, almost one-fifth of total gross domestic product, almost exactly the size of the revenues absorbed by the entire federal gov-ernment. (Gordon 1996, p. 4)

These 17.3 million nonproduction and supervisory workers almost equaled the number of people employed in local, state, and federal gov-ernment combined. Just counting the numbers of nonproduction and su-pervisory workers understates their importance since they tend to earn more than most workers. In fact, the $1.3 trillion that they earned equals almost one-fourth of all national income received by all income recipients. Twenty cents of every dollar we paid for goods and services went to cover the salaries and benefits of supervisory employees.

Think back to our discussion of unproductive labor in considering the magnitude of these costs. Even so, these figures underestimate the cost of supervision because they do not include the support workers who aid them nor the cost of their supplies (Gordon 1996, p. 35).

Are these costs not part of the necessary expenses of running business? Of course, some supervision is necessary, but not nearly the amount that looks over the shoulders of the typical worker in the United States.

Gordon noted that the data from the International Labour Organisa-tion, *Yearbook of Labour Statistics, 1994,* indicate that the United States had 13.0 percent of its nonfarm employees (including government) in man-agerial and administrative positions. In comparison, while Canada had al-most the same amount, 12.9 percent, Norway was next with a mere 6.8 percent. Belgium and Denmark followed with 4.5 percent. Other nations had even lower shares, including the Netherlands with 4.3 percent, Japan with 4.2 percent, Germany with 3.9 percent, and Sweden with 2.6. Data were not available for France, Italy, and the United Kingdom (Gordon 1996, p. 43).

Might the higher level of supervision in the United States just reflect the kind of industry found in this nation? Apparently not. Gordon found that the differences in the level seemed to remain when he compared individual industries rather than the total economy (Gordon 1996, p. 45).

Intensive monitoring can be self-defeating, encouraging perfunctory, rather than cooperative, behavior (see Moschandreas 1997, p. 42). Managers with an authoritarian mindset might well be inclined to interpret the resulting poor performance as evidence of a need for still more monitoring and supervision, further eroding productivity.

This high level of supervision is not a permanent feature of the United States economy. Instead, we should see it as part of a rapidly expanding move to ever more supervision. For example, in 1973 supervisory employees earned 16.2 percent of the national income of the United States. By 1993, these top-level employees had seen their share of national income soar to 24.1 percent. Since total employee compensation as a share of national income scarcely budged, from 56.6 percent in 1973 to 58.6 percent in 1993, the gains to the supervisors came from the hides of the supervised, either in the form of lower wages or reduced employment opportunities (Gordon 1996, p. 81).

Gordon pointed to an incident that suggests the emphasis that management in the United States put on its powers of command and control. In the late 1970s a group of 21 experts on corporate management convened to discuss the problems United States corporations faced in the marketplace. When presented with the statement, "In many cases control and power are more important to managers than profits or productivity," all of these experts agreed or agreed strongly (Gordon 1996, p. 75; citing O'Toole 1981, p. 115). Gordon went one step further, suggesting that this stress on command and control was not confined to the workplace.

To bring this point home, Gordon provided a scatter diagram for the incarceration rate in 1992–1993 and the share of administrative and managerial employees in 1989 for ten advanced economies. He excluded the United States since its incarceration rate is so extreme. Canada, Australia, and the United Kingdom formed one cluster with both high administration and high incarceration. Japan, the Netherlands, Denmark, Finland, Belgium, Germany, and Sweden formed another cluster of low adminstration and low incarceration, indicating a strong relationship between the incarceration rate and authoritarian methods of management. Gordon speculated that "habits of control bred in one social domain [such as control over the workplace] spill over to other areas of social life [such as the criminal justice system]" (Gordon 1996, p. 143).

Gordon's thesis reinforces the earlier association between crime and inequality. In a society where people are unequal, the more successful strata of society will find more justification for strong measures for control, whether they be more intense supervision at the workplace or more severe criminal sanctions. With fewer opportunities to fulfill their potential, those who are less fortunate are more likely to be resentful and frustrated, increasing the tendency for illegal activity. Hardly a recipe for a successful community.

The First Hint of Passionate Labor

Just what might substitute for the intense supervision that business imposes on its workers? In fact, workers who might not be inclined to work harder for monetary rewards or even in response to coercive workplace practices sometimes exert themselves as a form of asserting their independence vis-à-vis management. For example, Michael Burawoy described how workers applied their creativity in a piecework machine shop that produced parts for truck engines. The workers used the incentive system to relieve their boredom by turning their work into a game, sometimes attempting to produce as much as 140 percent of the norm for a relatively brief period of time. They did not produce any more than 140 percent because of fear that the norm would be revised upward (Burawoy 1979, p. 89).

Within this game, these workers developed a complex equilibrium with inspectors. If an inspector challenged them by rejecting articles with marginal tolerance, workers could retaliate by turning out scrap after the first piece was rejected, making the inspector look bad. Underlying their game was a need to relieve boredom rather than any direct economic motives (Burawoy 1979, p. 89). Michael Burawoy charged: "Political economy has conspired in a separation of economics and politics, never attempting to theorize a politics of production. Production has both ideological and political effects" (Burawoy 1979, p. 7). Burawoy described the workplace as an "internal state" (Burawoy 1979, p. 11). In the course of meeting challenges to management's authority, such as Burawoy described, enormous productive energies are dissipated. If society could tap into this creativity, the productive potential would increase by leaps and bounds. I will return to this subject in the final chapter of the book concerning passionate labor.

CHAPTER 6

The Waste of Human Potential

Racism

While we imagine our society to be a meritocracy in which the most talented people rise to the top, in truth, our society instinctively manages to detect superior abilities, whether real or imaginary, among the more privileged members of society. In fact, our society seems to regard the less privileged people merely as economic cannon fodder when the economy needs their service. The current system of management callously disposes of people just as soon as they are no longer needed. Possibly no group has experienced as shabby treatment as the descendants of those people who were involuntarily brought to this country as slaves.

In this environment, incarceration is a convenient tool. It reduces the population of unskilled labor for whom business has no need at the moment. It stands as a warning to those outside of prison that they must accept the discipline of the market. Finally, it offers lucrative opportunity for business investment, in building and sometimes even in running prisons (Parenti 1999).

Nationally, one in three black men in their twenties is imprisoned, or on probation, or on parole, largely as a result of the war on drugs (Butterfield 1995). In our central cities, the situation is even more chilling. For example, Jerome Miller's chilling book, *Search and Destroy: African-American Males in the Criminal Justice System* reports:

In 1992, the National Center on Institutions and Alternatives (NCIA) conducted a survey of young African-American males in Washington, D.C.'s justice system. It found that on an average day in 1991, more than four in ten (42 percent) of all the 18–35-year-old African-American males who lived in the District of Columbia were in jail, in prison, on probation/parole, out on bond, or being sought on arrest warrants. On the basis of this "one-day" count, it was estimated that approximately 75 percent of all the 18-year-old African-American males in the city could look forward to being arrested and jailed at least once before reaching the age of 35. The lifetime risk probably hovered somewhere between 80 percent and 90 percent. (Miller 1996, p. 7)

Miller went on to write:

In fact, most of the frenetic law enforcement in the black community had nothing to do with violent crime. . . . Sustained and increasingly technologically sophisticated law-enforcement intrusion into the homes and lives of urban African-American families for mostly minor reasons has left the inner cities with a classic situation of social iatrogenesis—a "treatment" that maims those it touches and exacerbates the very pathologies which lie at the root of crime. (Miller 1996, p. 9)

This experience with the law will limit the future employability of many of these young people. To make matters worse, the popular media convey images that tarnish virtually all Black youth with the stigma of criminality.

This environment creates what Gunnar Myrdal recognized as a "vicious circle" within which, he says: "Discrimination breeds discrimination" (Myrdal 1962, p. 381). For example, the employers often carry a sense of fear and mistrust of their Black employees. They tend to be quick to expect the worst of them. If that were not bad enough, employees often become inclined to look for minute signs that confirm their prejudice. In the face of such "evidence" employers often discipline their Black workers more intensively than their White counterparts. Anticipating such unfair treatment on the job, Blacks sometimes will slight their work rather than face the disappointment of failure in an effort to which they have devoted their heart and soul. Here again, the cycle of racism reinforces itself.

Environmental racism creates a different vicious circle. Polluters send their toxins to locations with the least political power. Sadly, the logic of economics confirms the rationality of this outcome. In the words of the previously cited memo of Lawrence Summers: "The measurement of the costs of health-impairing pollution depends on the forgone earnings from increased morbidity and mortality" (cited in Anon 1992). Since the poor, by definition, have less income to lose, as well as less power, the economy disposes of an inordinate share of contaminants close to the dwelling places of the poor, especially in communities of color.

Evidence is mounting that these toxins threaten people's health. Children raised in such environments are more likely to suffer learning disabilities as well as a broad array of physical ailments. For example, asthma is rampant in poor communities. Such effects will obviously be detrimental to their ability to perform in the workplace, ensuring that they will be more likely to remain poor and to raise their offspring in impoverished conditions.

Basketball, Racism, and Computer Technology

Racism is pervasive in the present day United States. Even in sports, one of the few venues where society associates Blacks with excellence, Blacks still face discrimination. For example, in cities where the population is more White, professional basketball teams hire fewer Black players (Brown, Spiro, and Keenan 1991). In an unpublished paper, Dan Rascher and Ha Hoang found that after adjusting for a number of factors, Black basketball players have a 36 percent higher chance of being cut than Whites of comparable ability. This statistic reinforces the widely held impression that while teams will want to employ the Black superstar for a better chance of winning, they will prefer a higher mix of White players on the bench to please their predominately white audience. A cynic might think of the employment of an excessive number of Whites in professional basketball as a form of affirmative action.

Today, in the United States, one aspect of racism is a stereotype of Blacks as natural basketball players rather than having a natural aptitude as engineers or business leaders. Accordingly, Blacks appear to have an unfair advantage over Whites in the sports arenas. I would like to explore this stereotype a little further.

True, many of the greatest basketball players today are Black. Why basketball? In the 1920s, the prevailing stereotype was that basketball was by its nature a Jewish game. According to the wisdom of the day, qualities such as sneakiness and guile, gave Jews a major edge that allowed them to be the best basketball players of the day.

I suspect that we will do well to steer clear of stereotypes and look for other influences on the social makeup of basketball players. The most commonsensical approach seems to be economic. After all, basketball is a very inexpensive recreation. You do not need elaborate facilities, such as certain water sports or ice hockey require. You do not even need the large open spaces that soccer or baseball requires. You can nail up a basketball hoop almost anywhere. So basketball is a wonderful sport for poor people, not because of the genetic makeup of Blacks, but because it is more available to people in the inner cities than, say, golf or polo.

Now let me shift gears for a second. A study by the National Telecommunications and Information Administration of the United States Commerce Department shows that at the end of 1997, 40.8 percent of non-Hispanic White households owned a personal computer, compared to 19.4 percent of Hispanic and 19.3 percent of African American households, a gap of 21.5 points (United States Department of Commerce National Telecommunications and Information Administration 1998).

Now, suppose that computers were as accessible as basketball hoops. This possibility is not nearly as farfetched as it might seem. After all, while basketball hoops may be readily available, the ratio of the cost of basketball shoes to the price of a computer continues to soar. Should this trend continue, in the near future, we might see innumerable young Blacks pounding away at their keyboards with all the exuberance that we now see on the basketball court. Those Black children with the advanced computer skills would enjoy the admiration of their peers.

Once excellence with computer skills becomes commonplace among the Black youth, some eminent scientists would no doubt set out to explain why the mental or biological makeup of Black children makes them ideally suited to computer programming. Perhaps we will hear that the result of natural selection over generations of cotton picking left them better suited than Whites to working with a keyboard.

Computer skills would soon command a lower wage, just as typing did, once the stereotype of operating a typewriter changed from being a man's job (since the typewriter was seen as a machine) to being a woman's job. In the wake of the depreciation of computer skills, opinion makers will bemoan the fact that so many Blacks waste their lives sitting in front of computers instead of following some higher calling where White youth seem to excel, perhaps such as basketball.

I want to emphasize the point that racism does not only harm Blacks. We all suffer from racism. Forget the moral and ethical implications of racism. That dimension of racism is so obvious that we have no need to subject it to detailed discussion. Instead, I want to insist that from a purely economic perspective, racism is a disaster for most people. Racism denies society the benefits of the talents of those people that racism stigmatizes. Racism is expensive for society. Educating people in schools and universities is far more economical than incarcerating them.

Yet, from another, more cynical standpoint, racism makes perfectly good sense—at least for certain privileged people. In a society rent with racism, people, both Black and White, are unlikely to look for the root cause of their difficulties. As a result, racism makes people infinitely easier to control. In this sense, we can view the divisiveness of racism as an effective glue of society insofar as disenfranchised people are less likely to challenge those who sit at the commanding heights of the economy.

The Waste of Talent and Creativity in General

Of course, the mere possession of a white skin does not guarantee great opportunity. While the waste of human potential may be more extreme among young Blacks, such waste is by no means confined to Blacks. In the United States, management predicates its basic philosophy on the proposition that labor should be as cheap as possible.

This seemingly reasonable objective—that labor should be as inexpensive as possible—leads to serious problems in the long run. In a market society, when any good is cheap enough, people will treat it with abandon. Labor is no exception in this regard.

The experience of the United States offers important evidence of the proposition that the expense and scarcity of labor is a great advantage to a society. By force of circumstance, labor was so scarce in the early years of the United States that wages were higher here than anywhere else in the world. I have shown elsewhere that by necessity business responded to the high level of wages with an intensive search for technology in an effort to economize on wages (Perelman 1993; and 1996).

Employers would not have felt so compelled to discover methods to economize on labor if wages had been less high. In effect, what business lost in the short run from the extra expense of higher wages they could recover in the long run from improved technologies.

Of course, these gains are not evenly distributed. Since the new technologies generally entail a larger scale of operation, some employers will find themselves squeezed out of the market; others will prosper mightily. Some highly skilled workers may suffer as machines displace their skills. In addition, some of the benefits from new technology will accrue to the purchasers of the product. Weighing all these costs and benefits together, most economists accept that the benefits of new technologies are positive. The one troubling aspect concerns the environment. Since labor-saving technologies frequently require substituting fossil fuels for human inputs, the environmental costs raise some serious questions.

In recent years, the United States has pursued policies designed to eliminate any increase in wages. This policy has been so successful that real hourly wages today are below the 1974 level, although some conservative economists dispute the methods for calculating wage rates. Not surprisingly, with the stagnation in wages, productivity growth in the United States has slowed dramatically.

In short, the experience of the United States confirms my assertion that devalued labor is wasted labor. Study after study shows that higher paid labor is more productive. The emphasis on low pay and low productivity—what I have called the Haitian road (Perelman 1993, Chapter 1)—is a dead end street.

Of course, raising wages is but a tiny step in the direction of creating an environment of passionate labor, but it is a step worth taking nonetheless.

The Tragedy of Modern University Education

Today's students get considerably less knowledge from their education than their predecessors did. In addition, the economy will take advantage of their education far less effectively. I attribute part of this problem to employers' continuing efforts to reduce the skill demanded on the job (deskilling) rather than either finding ways to upgrade the skill of their employees or, what is even more important, seeking to discover the latent skills of these workers.

At the same time, employers are continually increasing the qualifications that they require from new employees. For example, more and more employers prefer to hire college-educated workers for jobs that do little to take advantage of the skills acquired in college. Even on assembly lines, workers' educational levels are dramatically rising, although their jobs are unrelated to their education. For instance, 26 percent of workers at the Chrysler Windsor assembly plant have some college education (Thurow 1998, p. 32).

Over the last decades, the wages of a typical college graduate have not improved much, although a select few professions earn astronomical salaries. As a result, many students realize that the content of much of their education may very well be meaningless to their careers. For a good many jobs, only the diploma counts.

In effect then, education, rather than preparing people for life, including life on the job, merely serves as an elaborate signal to prospective employers. Just as a male peacock displays delicate tail feathers to attract potential mates, students must accumulate academic credentials in order to woo desirable employers, realizing that the content of their education will probably have every bit as little to do with the job as the peacock feathers would help a female peahen raise her brood.

Consequently, many students walk through their colleges and universities lackadaisically. I do not mean that they are lazy—merely uninspired. Many students are far from lazy, working 10, 20, or even 30 hours a week in order to support themselves. As a college teacher, I am convinced that the single greatest factor in the declining effectiveness of college education in the United States is the students' need to work long hours to cover living costs.

In a sense, the deficiencies of university training do not matter much, since corporations do a poor job of taking advantage of what the students do learn at the universities. Their recruiters take the course of least resistance and concentrate on the elite universities when they are looking to hire for the most desirable jobs, even if the most qualified workers might happen to be at less prestigious institutions. The name on the diploma, like the tail feathers of the peacock, is what counts. Ironically, many, if not most, employers are quick to tell the freshly minted graduates of such programs that they do not care what their professors taught them; they are expected to do exactly what they are told, regardless of what they learned in school.

Worse yet, the best potential employees may have been steered away from higher education altogether. Many students, especially from those poorer homes, never receive any encouragement for their education. They go to high school under unfavorable conditions, often finding discouraging signals at every turn. In addition, a good number of potential students have too many obligations to allow them to afford to go to school, even if they were to finance their education through part-time work.

In earlier years, society seemed to accept that only the most privileged young people would get the opportunity to enjoy higher education. For them, their peacock feathers were not for the benefit of prospective employers. In fact, such people rarely had to seek employment at all. At the time, higher education was, more than anything else, evidence of an intention to live a life in which the demands of work would be completely absent.

Sidney Smith, a witty clergyman who was a founder and frequent contributor to the *Edinburgh Review*, the most popular outlet for political economists at the time, maintained that:

> an infinite quantity of talent is annually destroyed in the Universities of England. . . . When an University has been doing useless things for a long time, it appears at first degrading to them to be useful. A set of lectures upon political economy would be discouraged in Oxford, probably despised, probably not permitted. To discuss the enclosure of commons, and to dwell upon

imports and exports, to come so near to common life, would seem undig-
nified and contemptible. (Smith 1809, pp. 50–51)

According to the wonderful Thorstein Veblen, whom we have met before,
higher education, especially the study of the "classics," was nothing more
than a form of conspicuous leisure characterized by "aversion to what is
merely useful" and "consuming the learner's time and effort in acquiring
knowledge which is of no use" (Veblen 1899, p. 255). In the same spirit,
the French sociologist, Pierre Bourdieu, referred to the modern educational experience as a form of cultural capital, with which the possessors
can lord their status above the more common folk (Bourdieu 1984, p. 80).

Already in Veblen's time another trend was afoot. With the rise of modern industry, economic forces demanded more trained engineers and scientists. Many educational institutions complied (Goldin and Katz 1999, pp.
38–39; and Noble 1979). Thanks to the efforts of higher education, the
output of these specialists soared. As a result, he noted: "Firms that had not
previously hired trained chemists and physicists did so at an increasing rate,
as did the federal and state governments. The number of chemists employed in the U.S. economy increased by more than six-fold between 1900
and 1940 and by more than three-fold as a share of the labor force; the
number of engineers increased by more than seven-fold over the same period" (Goldin and Katz 1999, p. 39).

Of course, a good number of prestigious colleges continued to concentrate on the traditional education of elites; however, many new colleges
sprang up as a higher education has become more commonplace. As a result, the share of the educational market controlled by the elite colleges fell
over time. Then, in the decades following the GI Bill, which is the subject
of the next section, millions of new students poured into higher education.
Today, more than one-quarter of young people attend colleges and universities at one time or another.

In the process, higher education has divided up along a rather obvious hierarchy, ranging from the elite universities to the junior colleges that largely
cater to working-class youth. For example, at Compton College in Los Angeles, the average family salary was $23,000. At East Los Angeles College, 60
percent had family incomes of less than $24,000 (Jacoby 1994, p. 21).

At nearby UCLA in 1991, over 60 percent of incoming freshmen at
UCLA had family incomes above $60,000 (Jacoby 1994, p. 21). UCLA is
a public university and not at the top tier of elite universities. At UCLA's
sister institution, the University of California, Berkeley, fully 20 percent of
the applicants to come from private schools. In fact, children from families

in the top 60 percent of the income distribution account for the entire rise in college enrollment during the past 20 years (Berdahl 1999).

In the more plebeian colleges and universities where most students study today, education has taken a decidedly more utilitarian gloss, with greater emphasis on vocational courses, especially business administration and computer technologies. The fare at the elite schools of today falls somewhere midway between Veblen's cynical vision of education and the more modern vocational studies that largely occupy less prestigious institutions. These elite schools often take great pride in their excellence in nonvocational education, although ancient languages and texts no longer occupy a place of honor in the curriculum (Winston 1999, p. 26). For example, Stanford University does not even offer an undergraduate program in business administration.

Nonetheless, students still gravitate toward more utilitarian areas of study, such as economics. Michael Lewis, an English major who went into investment banking reported "a strange surge in the study of economics" during his undergraduate years:

> At Princeton, in my senior year, for the first time in the history of the school, economics became the single most popular area of concentration. And the more people studied economics, the more an economics degree became a requirement for a job on Wall Street. There was a good reason for this. Economics satisfied the two most basic needs of investment bankers. First, bankers wanted practical people, willing to subordinate their education to their careers. Economics seemed designed as a sifting device. Economics was practical. It got people jobs. And it did this because it demonstrated that they were among the most fervent believers in the primacy of economic life. . . . Economics allowed investment bankers directly to compare the academic records of their recruits. The only inexplicable part of the process was that economic theory . . . served almost no function in an investment bank. (Lewis 1989, p. 24)

Lewis reported that 40 percent of the 1,300 members of Yale's graduating class of 1986 applied to a single investment bank, First Boston (Lewis 1989, p. 24; see also Summers and Summers 1989, p. 270).

In effect, these elite institutions carry out an elaborate charade. They pretend to offer non-vocational education, far superior to the trade school education that most colleges offer. In reality, these elite universities are merely providing their students with the cachet of an elite education, while actually paving the way for a lucrative career. These efforts in certifying their students for highly paid positions enhances the market value of their product, expanding parents' willingness to pay and ensuring a future flow of endowments (Winston 1999). At the same time, the less fortunate students consigned to more lowly institutions are left to accept employment that often has even less to do with their education.

I do not mean to denigrate either classical education or technical education. My only concern here is to point out the wasted educational efforts that plague higher education. To the extent that the organization of business creates a burden on society, devoting education to learning to manipulate business does not seem to serve much of a social purpose. To the extent that students feel constrained to take either classical or vocational courses merely to suit the fancy of employers, a great deal of time and effort is lost. Finally, to the extent that higher education fails to expose students to critical skills, and perhaps even more, to the extent that students' curiosity is left untouched because so much useless learning is imposed upon them, further waste ensues.

Once education succeeds in igniting students' curiosity, so long as they have access to adequate library resources and contact with interested faculty and other like-minded students, the educational process will work magnificently. When students must go through the motions of learning merely to acquire certification, the educational system will only succeed infrequently and accidentally.

The Lesson of the GI Bill

Although some schools such as the City University of New York welcomed the children of the working classes, the university environment was usually foreign to such children. Despite the expansion of both the state universities and the more widespread availability of technical training in higher education, prior to World War II, colleges and universities were still largely finishing schools for the children of the elite.

The GI Bill, which funded university education for about one-half of the surviving veterans following World War II, broke with that tradition (Skocpol 1998, p. 96). In the process, the GI Bill represented one of the most massive transformations of social capabilities in the United States.

Not everybody applauded this policy. Robert Maynard Hutchins, president of the University of Chicago, who was one of the great visionaries of higher education at the time and who had a well-deserved reputation as a liberal, dreaded the prospect of swarms of veterans entering into the hallowed halls of academia. He warned that "colleges and universities will find themselves converted into educational hobo jungles" (Hutchins 1944; cited in Olson 1974, p. 33).

Hutchins's apprehensions probably seemed well grounded at the time. After all, many of the returning veterans were not born into the aristocratic strata of the population that typically populated the elite colleges and universities, such as his own University of Chicago. Besides, a good number of these veterans had just finished participating in a violent conflict. That experience would not seem to be the appropriate training for aspiring college students. Hutchins probably realized that many of the veterans would be suffering from what we now call post-traumatic stress disorder, perhaps threatening the tranquility of the cloistered environment of a major university. Most important, perhaps, Hutchins dreaded the prospect of colleges and universities turning into vocational schools (Olson 1974, pp. 33–34).

In the end, all but the last of Hutchins' fears proved to be unfounded. The veterans by and large were far more serious about their studies than the typical well-bred, young college student. Judging from what I observed as a teacher during the Vietnam era, these enthusiastic veterans probably pushed many of the younger students well beyond what they would have otherwise achieved. After graduation, many of these veterans rose to positions that might have seemed unimaginable before the war.

We get a feel for the profound importance of the GI Bill for lower-class citizens from the account of a reunion of the 1944 high school class from Turtle Creek, Pennsylvania, a poor working-class community. The author, Edwin Kiester, Jr., himself a beneficiary of the GI Bill, wrote that his class had 103 male graduates in a high school class of 270. Kiester reported with some evident pride that:

> thirty earned college degrees, nearly ten times as many as had in the past; 28 of the 30 attended college under the GI Bill of Rights. The class produced ten engineers, a psychologist, a microbiologist, an entomologist, two physicists, a teacher-principal, three professors, a social worker, a pharmacist, several entrepreneurs, a stockbroker and a journalist [Kiester himself]. The next year's class matched the 30-percent college attendance almost exactly. The 110 male graduates of 1945 included a federal appellate judge and three lawyers, another stockbroker, a personnel counselor, and another wave of teachers and engineers. For almost all of them, their college diploma was a family first. Some of their parents had not completed elementary school—a few could not read or write English. (Kiester 1994, p. 132)

The experience of the Turtle Creek students was replicated throughout the country. As Kiester noted:

the first GI Bill turned out 450,000 engineers, 240,000 accountants, 238,000 teachers, 91,000 scientists, 67,000 doctors, 22,000 dentists, 17,000 writers and editors, and thousands of other professionals. Colleges that had languished during the Depression swiftly doubled and tripled in enrollment. More students signed up for engineering at the University of Pittsburgh in 1948 (70 percent of them veterans) than had in five years combined during the 1930s. By 1960 there were a thousand GI Bill-educated vets listed in Who's Who. (Kiester 1994, p. 130)

Nobody, to my knowledge, certainly no economist, has ever tried to take account of the full impact of the GI Bill, either for people such as Kiester's classmates or for the nation as a whole. Such a work would be daunting, to say the least, because the ramifications of this transformation are so extensive. Of course, the impact of the GI Bill goes far beyond the terrain that economists typically navigate.

Thomas Lemieux, a Canadian economist, and David Card, a fellow Canadian who teaches at the University of California, Berkeley and a recipient of the John Bates Clark award from the American Economic Association, have studied the Canadian version of the GI Bill, although from a relatively narrow perspective. The Canadian law did not affect Quebec as much as the rest of Canada since the French-speaking universities made no provision for returning veterans. By comparing labor productivity in Quebec and Ontario, they were able to get an estimate of the effect of the Canadian GI Bill on labor productivity. As would be expected, they found that productivity rose considerably faster in Ontario than Quebec.

This measure certainly understates the effect of the Canadian GI Bill in part because their methodology assumes that the improvements in Ontario would not affect Quebec. Certainly, some of the productivity improvements in Ontario would have filtered into Quebec, either because workers moved from one province to another or because of the spread of technology developed in Ontario.

In any case, we should not confine our interpretation of the GI Bill to matters of labor productivity alone. The GI Bill did much more than just increase labor productivity. It promoted a more egalitarian society by offering opportunities to people who would not otherwise have enjoyed them. To this extent, it promoted a sense of justice, as well as all the favorable outcomes associated with a sense of justice. In this sense, it may well have contributed to the spurt in productivity that the industrialized countries enjoyed, beginning a decade or so into the postwar period after the beneficiaries of the GI bill had time to rise to strategic positions.

Of course, a new GI Bill, or even universal access to education, would go only a small way toward what we will later describe as a regime of passionate labor. It would allow some people to come closer to realizing their potential. However, a GI Bill would not directly aid in overcoming the stifling, hierarchical relationships that define most jobs. To bring about a new system in which passionate labor would become the norm rather than the exception would require going far beyond merely offering more educational opportunities.

The Waste of Scientific Talent

If education were the labor bottleneck in our society, we would see employers scrambling to recruit as many educated people as possible. In fact, even many of the most highly educated people in our society have great difficulty in finding employment commensurate with their skills.

The Commission on Professionals in Science and Technology, a Participating Organization of the American Association for the Advancement of Science, found that most recent graduates of doctoral programs in science and technology are employed. Unfortunately, relatively few found employment in positions that could make full use of their training or expertise (Commission on Professionals in Science and Technology 1998). The vast majority of scientists with new Ph.D.s who obtain jobs in an academic setting must settle for temporary positions called postdoctorals that allow the graduate time to publish and gain other forms of distinction. A 1998 survey found that in fields such as biochemistry, molecular biology, chemistry, micro-biology, earth and space science, and physics, more than half of the recent graduates held temporary positions, even though the United States economy was experiencing a boom. Even in mathematics, a subject upon which much of modern technology depends, almost half of the recent graduates were in temporary positions.

Supposedly, the postdoctoral appointment will allow the graduate a better chance to obtain a permanent position, but the growing number of postdoctoral personnel suggests great competition for more secure employment (Stephan 1996, p. 1214). Today scientists who are unable to get permanent jobs just move from one postdoctoral position to another. In the physical sciences, for example, in the late 1970s as many as one out of every ten Ph.D.s who had been out for four years had a postdoctorate appointment; by the late 1980s, this number had grown to one out of every eight (Stephan and Levin 1992, p. 96).

The Commission on Professionals in Science and Technology survey discovered that only 13 percent of those who held their first postdoctoral appointment in 1993 were in tenure-track academic positions two years later. Only 16 percent of those in their second or more postdoctoral appointments in 1993 were on the tenure track in 1995. Three or four years after earning doctorates in biological science, 39 percent of the Ph.D.s remained in postdoctoral positions. Five to six years after receiving their degrees, 18 percent were still in these temporary jobs (see Commission on Professionals in Science and Technology 1998). Of course, the situation is far worse for those who earn their doctorates in the liberal arts, but we shall focus only on the sciences for now (Nelson and Berube 1994).

These postdoctoral scientists offer a reservoir of cheap labor for universities that can parlay the work of their underpaid researchers into lucrative grants from corporations or the government. In a rational society, low-wage exploitation of scientists makes no sense whatsoever. We will return to this thought in the next section.

To make matters worse, corporations are making strategic alliances with universities. For example, the department of Plants and Microbial Biology has entered into a five-year $25 million strategic alliance with Novartis Corporation. According to the agreement, Corporation representatives will hold two of the five seats on the committee that allocates research money for the department (Blumenstyk 1998).

In a sense, those who work in a corporate/academic department, even those who hold postdoctorate positions, might count themselves fortunate. At least, they have a tentative connection with the university research community. By 1987, 60.5 percent of employed Ph.D.s who had earned their degrees in the physical sciences between 1980 and 1982 were working in business or industry (Stephan and Levin 1992, p. 97). Many of these jobs have little promise of producing any social benefit. For example, a steady stream of Ph.D. physicists find work on Wall Street calculating strategies for investing in derivatives and other financial instruments (Mukerjee 1994). Still more, even if they do find work related to their education, they will often find their skills severely underemployed.

Even those who do find industrial employment in their fields of expertise may find their talents sorely underutilized because of the corporation mindset that discourages individual creativity. Consider the admission of Daniel P. Barnard, research coordinator of Standard Oil of Indiana, "We employ many people who, if left to their own devices, might not be research-minded. In other words, we hire people to be curious as a group. . . . We are undertaking to *create* research capability by the sheer pressure of money" (cited in Jewkes, Sawyers, and Stillerman 1958, p. 180).

True, the above statement was made in 1958, but I suspect that the situation has deteriorated since then. For example, at the time, the most important private employers of scientific talent, such as the research arms of AT&T and IBM, enjoyed monopolistic positions that gave them the leeway to give enormous degrees of freedom to their scientific talent. The creativity of these scientists led to key breakthroughs in our modern technology. In today's more competitive environment, these companies are no longer willing to offer the same sort of freedom. They demand research that is more focused on immediate problems.

Let me point to another ominous sign. Manpower Inc., the largest temporary employment agency in the United States, plans to provide holders of advanced physics degrees to corporate clients, mostly in the computer and electronics industries. Though few of the temps would be working directly as physicists, Manpower hopes to place them in related jobs, such as developing new computer chips or writing software programs. Mitchell Fromstein, Manpower's chairman, said that if Manpower's physicists catch on, "we'll offer chemistry Ph.D.s next." "Times have changed for physicists: There are jobs, but physicists have to be more flexible than they were in the past," said John Rigden, director of physics programs at the American Institute of Physics (Zachary 1996).

Yes, times are changing. As government continues to cut funds for higher education, the squeeze on scientists will no doubt become even worse, further discouraging young people from seeking careers in science in the future. As this process continues, the average age of university scientists continues to grow (Stephan and Levin 1992, p. 6). This aging of the scientific community represents another serious dimension to the crisis since most of the breakthroughs in science come from the young (Stephan and Levin, Chapters 3 and 4).

CHAPTER 7

The Waste of Doing
Business as Usual

Corporate Waste

The very mention of the term, bureaucracy, evokes images of immense waste. Although our modern corporate-owned media have conditioned the public to associate bureaucracy with government, in truth, the corporate bureaucracy far exceeds that of the government. Recall David Gordon's comparison of the size of supervisory labor with government employment. Add to that number the rest of the bureaucracy engaged in sales, marketing, administration, accounting, etc., and you will see that corporate government is far more extensive than that of public governments.

The media would have us believe that corporate bureaucracy is infinitely more efficient than government bureaucracy, even though the leadership of the government bureaucracy generally consists of the very same people that run the corporate bureaucracy. Business leaders often become indignant about wasteful practices on the part of government, yet remain quite tolerant of corporate waste when it suits them.

For example, the Reagan administration appointed the Grace Commission, led by J. Peter Grace, to lead a ruthless assault on federal waste, perks, and dubious spending. Mr. Grace took to his task with gusto, calling for the wholesale elimination of program after program all in the name of efficiency. Some time later, a *Wall Street Journal* reporter noted with irony, "Some say J. Peter Grace's company could use the same advice [that Mr. Grace gave to the government]." Apparently, his son, Peter Grace III used about $1.3 million from a Grace subsidiary for working capital without proper authorization. Among the beneficiaries of the corporate generosity was J. Peter Grace, who continued ro enjoy lavish benefits. The elder Grace was getting $165,000 for nursing home care, $200,000 for security guards, and $74,000 for a New York apartment for his family's use, as well as a full time cook from corporate funds. "The very things the Grace Commission said about the government were true within Grace," says Jack Shelton, an ex-employee and head of American Breeders Service, a cattle-breeding business Grace sold last year. "The corporate culture had gotten mired down, lazy and fat" (Miller 1995).

Henry Ford II inspired a stockholders revolt with his lavish spending. He equipped his office with $250,000 sauna, a private gym, a full-time masseur, a private dining room, and a Swiss chef. Each lunch cost $200 per person. He was also accused of having five to six employees just to tend his girlfriend's lawn (Cowling 1982, p. 87).

Supposedly, such corporate waste is virtually impossible—at least within the context of economic theory. Any economists worth their salt can easily go to the blackboard and "prove" why markets are efficient allocators of resources. Textbooks write in glowing terms about this supposed "allocative efficiency" of markets—the ability to direct resources to the activities where they produce the greatest possible value. The sheer force of competition will ensure that business will have no choice but to stamp out waste.

Like all economic theories, this one also makes strong assumptions, some of which are highly unrealistic. Nested among these assumptions is the unlikely supposition that everyone is actively trying to maximize profits.

Harvey Leibenstein took exception to the comfortable assumption of allocative efficiency. He introduced the concept of x-inefficiency, to differentiate his analysis of inefficiency from that of the typical economist, who accepts the dogma of allocative efficiency (Leibenstein 1966). The "x" in Leibenstein's x-inefficiency was to signify that something unmeasurable was at work—an unknown x-factor. Leibenstein's main evidence for x-inefficiency was his discovery of wide productivity differentials in nearby plants using similar technology.

How could inefficient plants survive if firms were engaged in a life-and-death struggle in which only the fittest would survive? In effect then, Leibenstein insisted that economists should realize that the supposedly all-powerful force of competition was relatively modest; that it did not require that business operate with anything like optimal efficiency. Instead, the economy allowed firms to enjoy a good measure of what the Nobel prize winning economist, Herbert Simon, termed "organizational slack" (Simon 1979, p. 509).

In other words, Leibenstein was saying that competition was not particularly effective in keeping the economy at peak efficiency. One author later referred to the respective losses from allocational inefficiencies, unemployment, and x-inefficiencies as "fleas, rabbits and elephants" (Vanek 1989, p. 93; cited in Schweickart 1996, p. 81).

Leibenstein was not alone in accusing corporate management of wasteful practices. According to Michael Jensen, a professor of economics at Harvard University, the highly paid managers who normally run large corporations have evolved into a collection of selfish bureaucrats who use corporate resources for their own aggrandizement (Jensen 1986; 1988; 1993; and Jensen and Meckling 1976).

Long before I had ever considered the nature of the modern corporation, I grew up near an intersection with a separate gas station on each of the four corners. My childhood friends and I used to wonder about the silliness of such an arrangement. We never saw all the pumps in use at any one time. Little did we know that decades later Professor Jensen would single out the petroleum companies as particularly outrageous examples of corporate waste.

Leibenstein's work reminds us of a simple truth probably known to everyone who has ever had a job. All organizations include a considerable amount of wasted effort. This waste can take innumerable forms. Supervisors may intentionally distort the system to benefit themselves in one way or another. In earlier times, when workers' organizations were stronger, business expressed outrage that labor could sometimes impose higher staffing requirements than business desired. While business was quick to denounce featherbedding, as this sort of supposedly unnecessary staffing was known, manager after manager gleefully added staff to his own operation in an effort to expand his prestige and importance within the firm.

Such x-inefficiency is most common in corporations that have acquired enough power to leave the organization relatively immune from the ravages of competition. This ability to do business without the inconvenience of competition is not permanent. Strong competition does occur, but only from time to time.

As I explained in an earlier book, *The Natural Instability of Markets* (Perelman 1999), these outbreaks of strong competition are associated with periods of depression or recession, which create catastrophic wastes of their own. So, in part, the level of x-inefficiency was extraordinarily high at the time Leibenstein was writing because the economy allowed firms to enjoy considerable "organizational slack." For example, David

Audretsch has reported that two decades passed before a third of the Fortune 500 would be replaced between 1950 and 1970.

One study compared the rate at which firms fell from the top 100 firms in the period, 1903–1919, with the period, 1919–1969. The rate of failures per 100 firms per year was at least 3 times as great in the earlier period. The author concluded: "The evidence reviewed above indicates that corporate capitalists had achieved a quite widespread and enduring consolidation of their positions by 1919" (Edwards 1975, p. 442). Another study found that turnover among the largest firms had already declined over the 1909–1929 period, just as we should expect after a period of corporate consolidation (Stonebreaker 1979). This "hardening of the industrial arteries and decreased competitiveness" of industry in the United States (Caves 1977, p. 40: and Caves 1980, p. 514) left considerable space for x-inefficiency.

As competition from imports heated up in the United States after the post–World War II boom cooled, the stability within the Fortune 500 declined. For example, one decade was enough for the replacement of a third of the firms between 1970 and 1980. The process continued to accelerate during the next decade, when a third of the firms would fall from the list every five years (Audretsch 1995, p. 7).

The FIRE Sector

Ironically, Professor Jensen's preferred remedy for wasteful corporate practices was to have new management teams acquire the management rights by way of corporate takeovers. Jensen believed that the new managers would have no choice but to be efficient, since the takeovers would saddle the corporation with huge debts. Management would have no choice but to pursue efficiency with the utmost diligence since any laxity on the part of management could mean bankruptcy.

This process created wastes of its own, including the human costs of slash and burn management. In addition, it helped to promote the rapid expansion in financial services. Christopher Niggle, an economist at the University of Redlands, illustrates the growing importance of the fire sector by analyzing the growth of financial services in the United States economy relative to other economic indicators. He notes that the ratio of the book value of financial institutions to the Gross National Product of the United States was 78.4 in 1960. In 1970, it was still only 82.9. By 1984, it reached 107.4 (Niggle 1988, p. 585).

Niggle suggests a second ratio to demonstrate the enormous growth of the financial sector: the ratio of financial institutions' assets to the assets of nonfinancial institutions. In 1960, this ratio was 0.957, meaning that the financial and the nonfinancial sectors were about equal. By 1970, the financial sector had overtaken the nonfinancial sector, boosting the ratio grew to 1.094. By 1983, the dominance of the nonfinancial sector had driven the ratio to 1.202 (Niggle 1988, p. 585).

Finally, Niggle reports the growing size of the part of the economy known by the acronym, FIRE, which stands for Finance, Insurance, and Real Estate. In 1960, the FIRE sector represented 14.3 percent of the Gross Domestic Product of the United States; in 1980, 15.1 percent. The 1983 share of the FIRE sector was 16.4 percent, meaning that within these mere three years the relative importance of the FIRE sector grew by more than it had in the previous 20 years (Niggle 1988, p. 585).

Just consider the explosion in transactions in the stock market. In 1960, 766 million shares were traded on the New York Stock Exchange. In 1987, 900 million shares changed hands in the average week. More shares were traded on the lowest volume day in 1987 than in any month in 1960. More shares were traded in the first 15 minutes of 19 and 20 October 1987 than in any week in 1960 (Summers and Summers 1989).

The stock market represents a relatively small share of all financial speculation. Speculators trade many different types of assets. For example, they buy and sell derivative securities, such as stock futures, which provide the rights to buy or sell stocks at a set price a specified time in the future. Organized markets in such derivative securities did not even exist in 1970. Today, the value of trades in stock futures exceeds that of the trades in stocks themselves. Trade in the New York Stock Exchange averages less than $10 billion per day; government bonds, $25 billion; daily trade in foreign exchange averages more than $25 billion. Trade in index options equals that of stock futures (Summers and Summers 1989).

How much do all these transactions cost? The combined receipts of firms on the New York Stock Exchange was $53 billion in 1987—an enormous sum considering that the total income for the entire corporate sector in the United States was only $310.4 billion. Beside these direct costs, corporations whose stock is traded in organized markets devote much time and energy in efforts to influence the markets. For example, chief executive officers of major corporations commonly spend a week or more each quarter just telling their corporate story to security analysts. In addition, both individuals and firms spend a great number of resources monitoring their portfolios, acquiring information about securities or making investment decisions. If these supplementary costs are one-half as much as direct payments to security firms, then the cost of operating the securities markets was greater than $75 billion (Summers and Summers 1989).

The FIRE sector produces no real goods. It does not even raise funds for business. In recent years, the corporate sector has bought more stock than it has sold (see Henwood 1997). For the most part, what the FIRE sector does is merely to shuffle assets around in order that speculators can bet against each other.

The Problem of Excess Entry

While an increase in the degree of competition might have the potential of limiting the extent of x-inefficiency, as I mentioned earlier, competition creates another sort of serious waste. Unlike the huge corporations where management feels secure enough to siphon off enormous resources, the small business sector is extraordinarily turbulent. The majority of the small start-ups are destined to fail.

For example, most of us have seen some location in a town where one shop after another fails. Each new venture opens up with some fanfare, seemingly expressing confidence that it will succeed where its numerous predecessors have faltered. One notorious location in my town comes to mind. A new restaurant appeared every few months over a period of several years.

While turnover among restaurants is notoriously high, most industries are turbulent. For example, more than 5 percent of all Canadian manufacturing firms closed down each year during the 1970s (Baldwin and Goreski 1991). Why then are so many investors willing to speculate when the failure rate is so high? Here we can turn to Adam Smith for some insight into this riddle. He observed:

> The over-weening conceit which the greater part of men have of their own abilities, is an antient evil remarked by the philosophers and moralists of all ages. Their absurd presumption in their own good fortune, has been less taken notice of. It is, however, if possible, still more universal. . . . The chance of gain is by every man more or less over-valued, and the chance of loss is by most men under-valued. (Smith 1776, I.x.b.26, pp. 124–25)

So Smith attributed the frequency of unsuccessful investments to a common character defect, which causes most people to overestimate their luck. He was not alone in his recognition of the tendency to overestimate good fortune. His contemporary, Samuel Johnson, also displayed an acute awareness of this all-too-human failing. James Boswell, Johnson's famous biographer, recounts a wonderful example of Johnson's insight. Thrale, the great brewer, had appointed Johnson one of his executors. In that capacity it became his duty to sell the business after Thrale's demise. When the sale was about to go on, Boswell reported:

> Johnson appeared bustling about, with an inkhorn and pen in his button-hole, likes an exciseman, and on being asked what he really considered to be the value of the property which was to be disposed of, answered—"We are not here to sell a parcel of vats and boilers, but the Potentiality of growing rich beyond the dreams of avarice." (Boswell 1934, vol. 6, pp. 85–86)

Of course, if people can only hope to succeed in business, then excessive entry should not be unexpected. When a new sort of business opportunity emerges, an excessive number of people rush into the field, soon creating an overcapacity.

Inertia and Turbulence

To some extent, turnover is desirable. How else could the economy eliminate hopelessly inefficient, and even wasteful, firms? Economic theory promotes the concept that competitive forces necessarily succeed in weeding out less efficient operations, while ensuring that the fittest survive. While many obviously inefficient firms do fall to competitive forces, convincing proof of the efficiency of competition is wanting.

Because large firms are generally unreceptive to new ideas, young firms provide the seedbed of the economy in which innovative ideas can take hold (Beesley and Hamilton 1984). Even previously innovative, small firms, once they mature, eventually tend to become blinded to good ideas. A chain of events beginning with Xerox represents a classic example:

> Chester Carlsson started Xerox after Kodak rejected his new idea to produce a copy machine, telling him that his copy machine would not earn very much money, and in any case, Kodak was in a different line of business. . . . Stephen Jobs started Apple Computer after this same Xerox turned Jobs away, telling him that they did not think a personal computer could earn very much money, and in any case, they were in a different business. (Audretsch and Acs 1994, p. 174; see also Audretsch 1995, p. 54)

Later, Apple, in turn, became relatively stodgy and stumbled because it failed to introduce very exciting products for a relatively long period of time, given the speed with which the computer industry evolves.

Although stodginess remains a serious economic problem, an economy in which firms do not enjoy a certain degree of stability imposes a different set of costs on the economy. In other words, in a turbulent economy in which survival is precarious for both large and small firms, business suffers enormous losses, creating widespread wastes in the process. To begin with, such an economy will dissipate huge amounts of time and energy in expensive start-up costs. In addition, the high risk of failure will discourage investment on the part of intelligent investors. Finally, because business failures disrupt people's lives, the human costs of turbulence is enormous.

For example, the *Journal of the American Medical Association* published a study of the effects of moving on children. It reported that frequent relocation was associated with higher levels of all measures of child dysfunction; 23 percent of children who moved frequently had repeated a grade, compared to 12 percent of children who never or infrequently moved. Eighteen percent of children who moved frequently had four or more behavioral problems compared to 7 percent of children who never or infrequently moved. Using statistical adjustments to eliminate the possible effect of other factors, the authors found that children who moved frequently were 77 percent more likely to be reported to have four or more behavioral problems and were 35 percent more likely to have failed a grade (Wood, Halfon, Scarlata, Newacheck, and Nessim 1993). Rarely do economists recognize the economic cost of turbulence (Mankiw and Whinston 1986).

Of course, we could not eliminate all turbulence from the economy. Indeed, we would not even want to do so. In effect, then, an economy should be fluid enough that inefficient firms will disappear while efficient firms should flourish. I know of no example of such an economy.

How can we evaluate the degree to which turbulence and inertia create waste in our economy? We lack the necessary data to go beyond an impressionistic estimate. My best guess is that we have too much inertia among the giant corporations and too much turbulence in the small business sector. The excess inertia leads to massive x-inefficiencies while the turnover among the small businesses consumes enormous quantities of resources and labor that could be put to better use.

The Billing Economy

The popular press often speaks of an information economy. I would prefer to think of the operative word as "billing." Let me explain.

In a typical firm, even an industrial firm, most of the employees do not directly produce anything. Instead, they just keep score. Amidst all this paperwork, people process and/or pay bills. Supervisors work hard to ensure that workers do not get paid too much for their labors. Others plot to sell some good or service in order to saddle another person or business with bills.

Think of your last trip to your doctor's office. Besides the doctors, I imagine that the office had a receptionist, perhaps one or more nurses, and a quite a few people working on billing and filling out forms. You may not have actually seen all of these other people because they might be in a separate room or they may have been working for another company that contracts with the medical practice. In any case, the health care professionals represent only a small part of the overall personnel in the health care industry. Instead, the billers, along with their counterparts in the HMOs or insurance companies, make up the bulk of our health care sector.

Hospitals in the United States employed 435,000 administrative staff and served 1.4 million patients in 1968. Although the average daily patient population dropped to 853,000 by 1992, administrative employment actually rose to 1.2 million—in large part because information processing consumed an increasing amount of staff time (Woolhandler, Himmelstein, and Lewontin 1993, p. 401). Symptomatic of this emphasis on billing, a recent federal government study on preparedness for the Y2K computer problem found: "About half of doctors, hospitals and nursing homes report that their billing and medical records computer systems have been fixed for Year 2000 operation, but less than a third said they had finished checking their biomedical equipment, a federal survey released yesterday showed" (Barr 1999).

The core of our economy now consists of people employed to send bills, pay them or collect them, together with those who attempt to induce us to incur new bills. Even many people who actually build things do so, at least in part, in order to facilitate the billing process. Think of the people working to produce paper products, or computers or the trucks to carry the bills, etc.

No wonder the key figure in our information age has the given name of Bill, who wants to position his company so that he can collect a portion of each bill in the Billing Economy. His family name, Gates, is no less appropriate. After all, those who collect the bills have to construct gates in order to exclude all those who do not pay the bills.

I confess that I too once served as a minor functionary, manning the gates. While in college, I worked taking tickets at a theater. I did not produce better entertainment; I just guarded the door lest someone dare to enter without paying.

Of course, any economy must devote some effort to keeping score; however, in an economy where people trust each other, the necessity of keeping score is far lower. For example, family members do not normally send out bills to each other demanding payment for performing routine household chores. They can handle such matters informally.

With trust, we no longer need to be as obsessive about people taking more than their fair share of the economic product. In the process, society would have more time and resources available for productive needs.

The unwarranted emphasis on the Billing Economy is symptomatic of a larger failing in our society. While we devote considerable energy to count the countable, we typically neglect the rest. Occasionally, some economists do make an attempt to quantify nonmarket work, such as caring for children or cooking. This obvious, but usually uncounted part of the economy may well exceed the part that falls within the purview of economics (see the estimates in Eisner 1985; 1988; and Scitovsky 1976, p. 87).

If the demand for peanuts falls or if profits in the video game industry sag, government commissions will rush to meet the crisis. All the while, serious deterioration in the uncounted sectors will escape public notice, unless, of course, such problems happen to threaten either property or profits.

The Metered Life

With the spread of the Billing Economy, corporations attempt to maximize their profits by debundling their products—that is, by separating a transaction into a number of separate transactions. For example, banks and phone companies used to charge a rate that covered a whole range of services. Today, they have separate charges for using an ATM machine or for requesting information. This procedure allows them to extract a maximum payment or to economize by reducing the services that they offer. While debundling serves the seller quite well, it can be irksome to the consumer.

Let me explain what I mean. Engaging in a transaction requires a certain effort. It can be distracting or it can take some time, just as when a toll booth requires drivers to wait in line to pay their fees. Let me turn to a less realistic example to make my point.

Ordinarily a renter pays a flat fee to use an apartment for a month. Suppose that the landlord greatly reduced the basic rent, but debundled the services so that the renter paid separate charges for ordinary uses of the apartment, such as accessing a closet or a cupboard or using the bathroom. The landlord could benefit from this arrangement by earning more rent from the harried or affluent renter who might not be bothered with such trivia.

A frugal renter could save money, but to do so might require constantly paying attention. For example, you could save money by making sure that you had everything with you before you took a shower so that you would not have to reenter the bathroom. Presumably, the landlord would gain from the frugal renter's behavior since it would cause less wear and tear on the apartment. While this debundled apartment might be less expensive for the frugal renter, it would be far less of a refuge from the world of work and commerce.

Within this metered world, a deal, such as a rental contract would be nonstandardized. The buyer would have to scrutinize every contract to make sure that all the minutiae would be covered, creating still another distraction. In effect then, the sellers are throwing additional costs onto the buyers, while creating more revenues or superior economies for themselves.

I believe that the current pattern of the billing economy is leading in the direction of ever more debundling. While apartment rents might not be debundled in the near future, other types of activities will be. Banks and phone companies have profited from debundling, racking up separate charges for specific elements of service, such as the use of an ATM machine or requesting a telephone number.

In this debundled, metered world, people will experience more hassles in order that corporations may profit. In what is perhaps a sign of the times to come, Coca-Cola has developed a beverage dispenser it has programmed to raise the cost of a drink when the ambient temperature increases. This technology is already in use in Japan (Hays 1999).

Some people favor the use of metering for reasons of efficiency rather than profit. For example, they propose congestion pricing for roads or peak-load pricing for electricity. Charging tolls during rush hours, for instance, would discourage driving at those times. If tolls were high enough, affluent people could drive home on roads largely unencumbered by the traffic created by less-fortunate citizens, who would presumably arrive at work early and leave late to save on the tolls—if appropriate mass transit were not available.

To the extent that these tolls would fund mass transit, this scheme might make some sense. However, as metering becomes more common, more and more people will be working to "keep score," making sure that nobody uses "too much" from the perspective of the market. In reality, metering means rationing by pricing the poor out of markets or making them pay so much that their standard of living will decline in other dimensions. As everything becomes metered, people will have to pay for using everything public and private. With the spread of this logic, the sort of public spaces where people could meet and relax without the oversight of a commercial monitor disappear.

In the metered world, people only appear as blips in profit and loss statements. Time is money. Too much time spent on a job is a waste of money. Meter it. Work time is a special sort of time that firms carefully ration. They regard outlays on wages as a drain on profits. Meter it.

Where the convenience, welfare, and even health of people are concerned, so long as they do not appear on the balance sheet, they are of no concern. In the metered world, time is money only in specific contexts. Other people's time is a free unmetered good for the corporation. Firms thoughtlessly invade private homes through telemarketing on the outside chance that someone will be lured to change a phone company or have a photograph taken. This type of marketing makes perfectly good sense. The time of the telemarketer is relatively cheap and the time of the potential customer is of no importance whatsoever.

Some time must be spent on potential customers to get their business. As they say, "the customer is king." In reality, the claim of consumer sovereignty is only a charade. The company knows that after the completion of the sale, the customer is liable to move on to a different establishment. As a result, time spent on the customer is often carefully rationed after the sale. As a result, corporations all too often treat the precious time of their previous customers as a free good, for instance, forcing them to waste countless hours of frustration while navigating voice-mail mazes.

Such inconveniences and annoyances imposed on people by the Billing Economy go unnoticed in official accounts of the economic efficiency. While economists chalk up the economies in reducing labor time on the job as efficiency, this form of waste will forever elude measurement.

Concluding Remark

The small sample of wastes and lost opportunities discussed in Part 1 suggests an enormous opportunity for improvement. Of course, no society could never achieve the ideal of eliminating all the wastes discussed above, let alone match the perfection of a Carnot engine by obliterating all waste. Nonetheless, society could go far beyond this ideal by transforming the way we organize our society.

Before I address the nature of this transformation in Part 3, I will discuss how other economists have touched upon the question of waste.

PART 2

Alternative Approaches to Analyzing Waste

CHAPTER 8

A Review of the Literature

Absence of Waste as an Economic Category

Economic theory purports to show that markets necessarily lead to efficient outcomes. Even abuses by business supposedly have little impact on the economy. For instance, an inordinately influential article by Arnold Harberger estimated that the waste imposed by monopoly amounted to a mere one-tenth of one percent of the Gross National Product (Harberger 1954).

Given that mindset, economics has rarely taken account of the issue of waste, except to the extent that specific parties refuse to conform to the dictates of the market. Perhaps the sole clear example came from the earlier discussion of John Stuart Mill, who, while professing sympathy for the working class, blamed the problem of waste on workers.

While economists almost universally ascribe efficiency to business, economists over the last couple of centuries have continued in Mill's footsteps, worrying about the habits of working-class people insofar as their behavior interfered with productivity. For example, just as Herbert Hoover worried about the effect of alcohol consumption on the productivity of workers, so too Irving Fisher, perhaps the most important American economist of the twentieth century, spent considerable time and energy trying to calculate the amount of productivity lost due to the consumption of alcohol and cigarettes. Even today, employers continue to make similar types of estimates (Winslow 1998).

This analysis of the question of alcohol consumption illustrates an interesting paradox about the nature of efficiency measurement in a market economy. Economics considers that all marketed goods make a positive contribution to the Gross National Product, which is intended, in turn, to be an indicator of social wellbeing. At the same time, from the perspective of Hoover or Fisher, the production of alcohol or tobacco is detrimental to the economy. However, the fault does not lie with the economic system as a whole, nor with those who produce, advertise, or market the product. The individual worker who consumes the alcohol supposedly bears the full responsibility.

During the formative years of economic theory, economists engaged in a discussion that did seem to indicate an interest in the concept of waste. At the time, some economists concerned themselves with the division of working people into the categories of productive and unproductive labor. At first glance this procedure seems to be similar to some of the discussion of waste found in the first chapter. In truth, the spirit of the discussion of unproductive labor was quite different from the analysis of waste I have been suggesting.

Adam Smith, a pioneer in the discussion of unproductive labor, was engaged in a political project rather than economic analysis. He considered all labor to be productive labor, no matter how it was employed, just so long as it was engaged in a money-wage relationship. For example, suppose that an aristocrat maintained a tailor at his estate. The tailor would get a living in the form of food, shelter, clothing, and other incidental items, but he would not receive a monetary wage. Smith would classify this tailor as an unproductive laborer, even though the same tailor, doing the same work as an employee engaged in producing garments for sale on the market, would be considered to be productive labor.

Smith's procedure for classifying productive and unproductive labor was, as I suggested earlier, above all else political. He intended to denigrate the roles of all people who were not yet linked up in the modern market economy while elevating all those who were so engaged. In short, Smith was merely trying to argue against all traditional forms of employment in favor of market relationships (see Perelman 2000, chapter 9).

Once the older economic forms had considerably shrunk relative to the market economy, the economists became critical of the category of unproductive labor, recognizing that this concept could be used to censure certain activities within the market society. Today, students of the history of economics look back at Smith's idea as a quaint idiosyncrasy without realizing its social context.

In effect, Smith's vision of unproductive labor continues today in a slightly different guise. Rather than attack those associated with the gentry as unproductive, contemporary economists categorize government activity as unproductive. Recall the earlier discussion of the theory of rent-seeking behavior, within which economists interpreted government regulation as a pure waste, while the benefits of regulation, such as pollution control passed unnoticed.

Ironically, this literature grew out of a dissatisfaction with Harberger's dismissal of the wasteful impact of business monopoly. While Harberger insisted that business itself did little if any harm, the theorists of rent-seeking saw an opportunity to tar representatives of governments from whom business regularly solicits favors, without casting aspersions on private enterprise (see Hines 1999).

Similarly, the manner in which the United States government treats public investment in its national accounts reflects another ungrounded bias in favor of private, profit-maximizing activity as opposed to public endeavors. The late Robert Eisner, who tirelessly worked to correct the misconception that the government statistics perpetuate, pointed out:

> If United Airlines buys a new plane—that is investment—If Chicago builds a new runway for that plane to land on, that outlay is considered "government expenditure" and is counted, implicitly if not explicitly, as consumption. Similarly, a new truck purchased by business is investment. The highway that is constructed for it to ride on—unless a rare private toll road—is not. If the Internal Revenue Service spends $100,000,000 for new computers to process tax returns, that is not investment. . . . If business firms buy new computers . . . that is investment. (Eisner 1994, p. 51)

In conclusion, economic theory has avoided taking on the concept of waste seriously—beyond using the concept to make ideological points against those whose behavior might not suit the interests of business. We have already seen how the conflictive nature of a market society creates an enormous burden of economic waste. When even an eminent economist, such as Harvey Leibenstein, dares to suggest that a market might give rise to inefficiencies, the watchdogs of orthodoxy subject him to harsh rebuke (see Perelman 1999, pp. 95–97). Most economists soon learn the limits of "legitimate" inquiries.

Persuasion

In a market society, people and organizations devote considerable time and effort to persuading others to conform to their wishes. Adam Smith raised this issue, but in a way that did not seem to discredit market society. In fact, he considered this behavior to be so ingrained that he considered it to be a core element of human nature. Smith wrote:

> If we should enquire into the principle in the human mind on which this disposition of trucking is founded, it is clearly the natural inclination every one has to persuade. . . . Men always endeavour to persuade others to be of their opinion even when the matter is of no consequence to them. If one advances any thing concerning China or the more distant moon which contradicts what you imagine to be true, you immediately try to persuade him to alter his opinion. And in this manner every one is practicing oratory on others thro the whole of his life. You are uneasy whenever one differs from you. (Smith 1978, p. 352)

For Smith, the instinctual inclination to persuasion was a universal force so powerful that it caused the spontaneous emergence of market society, except where it was forcefully repressed. Following Smith's logic, attempts to persuade might have some negative consequences in some instances, but Smith seems to have implicitly accepted that persuasion in the commercial sphere necessarily leads to a preponderance of positive outcomes. The more persuasion, the better.

Presumably, rationality will prevent people from letting another person's persuasive skills manipulate them into irrational behavior, at least in the sphere of commerce. Smith, however, was not altogether consistent in his discussion of rationality. Recall his idea that people who embark on new business ventures suffer from an "over-weening conceit" that makes them overconfident. To prevent them from persuading lenders to succumb to their entreaties, Smith recommended laws to keep interest rates low. By this means, he hoped that lenders would be more careful in sorting out the ill-conceived proposals from the solid ones.

Smith's treatment of persuasion won him a more respectful reading than his analysis of unproductive labor. Recently, Donald McCloskey and Arjo Klamer followed Adam Smith's lead in treating persuasion as an integral part of the economy. They produced an intriguing effort to get a rough handle on the extent of persuasion in the economy. Although they were not directly concerned with the subject of unproductive labor and wasted effort, their work has some parallels with the approach of this book. However, they, like Smith, seem to treat persuasion as an effort to convey rational thought.

To estimating the extent of persuasion in the economy, they made rough guesses about the degree of persuasion associated with various occupations. They counted the work of actors, people in public relations, and others as 100 percent persuasion. They credited the bulk of the persuaders, such as editors, reporters, and sales people, with 75 percent persuasion. Finally, they counted another group, including police and nurses, with 25 percent persuasion (McCloskey and Klamer 1995, pp. 192–3). They came up with an estimate that one-quarter of the Gross Domestic Product of the United States consists of persuasion.

McCloskey and Klamer's idea of persuasion is admittedly idiosyncratic. They assume that all persuasion, like alcohol, makes a positive contribution to the economy. Of course, the persuader need not have the persuadee's best interest at heart.

Neither Smith nor McCloskey and Klamer seem to acknowledge the potential confusion that persuasion can create when people's interests are conflicting. Suppose that I come to you with a computer problem. If you make your living by selling computers, you might want to persuade me to buy an unnecessary upgrade to correct my problem. Your persuasion might cause me to spend money for resources that I do not need. My purchase will indeed add to the Gross National Product, even though it may not make any contribution to my welfare. In addition, the time and effort, as well as resources, consumed in attempting to persuade others can represent waste, especially if the object is to persuade others to act against their own interests. Once we drop that assumption and recognize that much persuasion represents economic waste, the McCloskey and Klamer analysis reinforces the theme of the first chapter of this book.

Keep in mind that persuasion, at least as I understand the word, typically occurs when a difference, or at least a potential difference exists between two parties. If I come to you for help, say, in helping me program my computer, you are not engaged in persuasion. You are helping me in a cooperative endeavor. I think that we would have to twist the notion of persuasion quite a bit to say that you are persuading me to operate my computer in the proper manner. Here again, I am violating the spirit of the McCloskey and Klamer analysis, which classifies such activities as persuasion, at least when they are carried on commercially.

So, we would have to recognize that McCloskey and Klamer do not acknowledge the distinction between the positive and negative effects of persuasion. Of course, we cannot easily classify specific activities as positive or negative persuasion.

Take advertising for example. While I consider most advertising to be an intrusion that clutters my life, some rare advertisement may conceivably provide me with objective information that would allow me to choose the best product for my needs. If advertising merely attempts to persuade me to purchase a product for which I have no need, it has done little to make the world a better place.

Advertising also indirectly destroys information. For example, in recent years, advertisers have successfully pressured many newspapers and magazines to tailor the content of their publications to cast a positive light on the products featured in the advertisements. Writers then have to fashion their materials in such a way that they can satisfy the advertisers, as well as the readers of the publication and the editors. The typical reader has no way to disentangle these motives. In addition, I should mention that such pressures lead publishers to spread misinformation, which often leads to social policies that are both economically and socially destructive.

Some forms of "persuasion" seem to offer virtually no useful information. Two examples come to mind. I have never heard of anybody being grateful for either the spam that clutters electronic mail accounts or the calls from telemarketers that inevitably come at an inopportune time. According to the Direct Marketing Association, direct marketing employment grew at a rate of 5.4 percent per year between 1993 and 1997, more than twice as much as total employment in the United States, which increased at an annual rate of 2.4 percent. As of 1999, about 5.3 million people were employed in such pursuits (Direct Marketing Association 1999).

Suppose that we had a way to make a distinction between efforts to manipulate people for an ulterior purpose and the act of providing information. Even here, we still could not clearly identify positive and negative persuasion. For example, if a nurse persuades you to take measures to be healthier, her or his efforts would seem to serve a positive purpose, at least until attempts at persuasion degenerate into a form of nagging. Unfortunately, the typical individual lacks the means of making that distinction between manipulation and the provision of information.

Only in rare cases can we identify who might and might not be objective. For example, most people are familiar with the enormous sums of money paid to scientists whose objective was to throw doubt upon the harmful effects of tobacco. For the most part, we will not go far astray if we assume that scientists who dismiss the health hazards of tobacco are financially dependent on the tobacco corporations. Other examples of scientific self-interest are less well-known. For example, Paul and Anne Ehrlich described how well-funded scientists and journalists disputed the dangers of environmental damage, just as cynically as the tobacco scientists who attempted to cloud the dangers involved in smoking (Ehrlich and Ehrlich 1996, pp. 36 ff).

Although the vast majority of scientists agree that our society is creating conditions that threaten to upset global climate patterns, the typical news broadcast finds the need to show both sides of the "debate," lest it offend present or potential advertisers. Each camp is allowed an equal opportunity to persuade the public. As a result, we are led to credit each side with an equally strong position.

Of course, the small numbers of the doubters is not, in itself, evidence that they are wrong. Small numbers of scientists have overthrown widely held misconceptions in the past. However, since the scientific minority is also financially dependent on a larger corporate interest, we do have reason to be skeptical of such minorities.

The example of corporate science illustrates how we might turn the idea of Klamer and McCloskey on its head, claiming that the huge efforts put into persuasion are an indicator of waste. Of course, such an assertion is an overstatement, but not nearly so much of an overstatement, I suspect, as it may seem. Still, I believe that we can take persuasion to be an index of social loss with as much justification as Klamer and McCloskey use it as a measure of a positive contribution to the economy.

While some of the persuaders do perform socially valuable services, others do so much damage that they more than cancel out the good efforts of the others. Consider the role of corporate science again. The efforts at persuasion by such scientists impose a double cost on society. First, many of the people employed by such corporate interests are highly skilled. By diverting their efforts to the production of misinformation, society loses the opportunity to take advantage of their talents as well as the resources that supported their work.

Second, most economists consider capital to be essential to a society's economic potential. In a modern economy, the capital stock of our information may be more important than the physical capital stock. For example, the German and Japanese people rebuilt their shattered economies at the end of the World War II because their technical and organization information was intact, notwithstanding horrendous misinformation associated with Nazis. Since scientific misinformation represents a net subtraction from our stock of information, the efforts of these researchers harm our economic potential.

Coase

Ronald Coase won a Nobel Prize in Economics largely for his influential article, "The Nature of the Firm" (1937). His work bears closely on our discussion of waste. Since Coase is very conservative, the essence of his article might seem very surprising. According to Coase, firms exist only because of deficiencies in the market. In Coase's words: "The main reason why it is profitable to establish a firm would seem to be that there is a cost of using the price mechanism. . . . The costs of negotiating and concluding a separate contract for each exchange transaction which takes place on a market must also be taken into account" (Coase 1937, p. 390). Just take a minute to think about what Coase meant. In an ideal market society with perfect efficiency, firms would not exist. You can imagine what such an economy would look like by picturing the disassembly of a firm into its component parts, into what the business press now calls a virtual firm. For example, the virtual firm can rent factories, offices, and equipment from individual owners rather than owning them. Instead of hiring people into long-term employment, it could subcontract with other firms to perform the work, or each individual worker could negotiate an independent contract, perhaps even daily, that would specify the content and remuneration for each unit of work.

For example, suppose that our firm manufactures automobiles. It could purchase every component from individual suppliers. It could even purchase partially assembled cars from individual suppliers. Our firm could contract with other firms to handle sales, advertising and other services. So, in the end, our disassembled firm would consist of an individual with a set of contracts with workers and suppliers of inputs.

After we similarly disassemble all the firms in our economy, the physical appearance of our world need not change. We would still see the same plant and equipment, as well as the same people doing the same jobs. Coase's insight was to recognize that this disassembled economy would be an inefficient economy.

Why? The owner of our virtual firm would have to negotiate satisfactorily the minutiae of millions of contracts. No one could have the foresight to anticipate every contingency. Even if someone would be able to succeed in effectively negotiating the contract, the cost of drawing up these contracts would be enormous. In all likelihood, the operation would find itself mired in a morass of confusion. Coase was specifically referring to this confusion as part of the "cost of using the price mechanism."

Coase labeled these inefficiencies "transactions costs." For Coase, the firm exists to minimize transactions costs. As firms grow, they also accumulate their own sort of bureaucratic inefficiencies. The balance between these two inefficiencies determines the size and shape of the firms.

Some of the problems that we have addressed in this book naturally fall under the rubric of transactions costs. Recall, for example, my discussion of the transactions costs associated with the creation of a metered society. In this sense, Coase's project and my own have something in common.

In another sense, however, this book and Coase are at odds. Coase is a highly ideological defender of business. He believes that the market is always superior to any government-imposed solution. He insists that the rise of bureaucratic inefficiencies ensures that socialism will necessarily be an inefficient system. Most important, he gives no indication that he has ever given any consideration to a cooperative, socially-determined outcome.

So, while Coase admits that transaction costs are a dead-weight loss to society, he is confident that the market economy will minimize these losses on its own. More important, from Coase's perspective, any attempt to interfere with the free functioning of the market will only cause transactions costs to multiply, weighing down the economy with unnecessary activities. He rules out the possibility that a public agency can minimize transaction costs.

As I am writing now, more than 50,000 acres of land are ablaze in Butte County, where I live. The usual grumbling about government waste is muted, while hard-working, courageous fire fighters working for the government risk life and limb, not to mention their lungs, to preserve beautiful houses in the hills. Nobody seems to be calling for privatization at this time.

David Landes, an economic historian from Harvard, relates a different arrangement for fighting fires:

> In Ottoman Turkey, firefighting was in the hand of private companies, who came running when the alarm sounded. They competed with one another and negotiated prices with house owners on the spot. As the negotiation proceeded, the fire burned higher and the stakes diminished. Or spread. Neighbors had an interest in contributing to the pot. 'Twixt meanness and greed, many a house fire turned into mass conflagration. (Landes 1998, p. 520)

I cannot speak for Ronald Coase, but I know which sort of fire protection I would prefer.

Williamson

Oliver Williamson offered a more nuanced analysis than either Coase or the approach suggested by McCloskey and Klamer. In the end, although Williamson's tone is not doctrinaire, he too is a strong defender of the existing system of hierarchy. After all, the message of his major book echoes Coase, proclaiming "that the economic institutions of capitalism have the main purpose of economizing on transactions costs" (Williamson 1985, p. 17).

In Williamson's world, people are always and everywhere ready to behave opportunistically. He referred to this phenomenon as "self-interest-seeking-with-guile" (Williamson 1984, p. 198). Such activity causes huge losses to society.

While Williamson went into great detail in analyzing the way people take advantage of employers or of those with whom they do business, he firmly rejected the parallel notion that "bosses exploit workers and hierarchy is the organizational device by which this result is accomplished" (Williamson 1980, p. 7; see also Williamson 1985, p. 261). However, he failed to back up this claim. As Gregory Dow has charged:

> [A]uthority at a high level is invoked as a means of restraining opportunism among subordinate agents. This illustrates a more general phenomenon: transaction costs theorists tend to see authority primarily as a remedy for opportunism, rather than as a device which might be abused in an opportunistic fashion . . .

This omission is puzzling, because transaction cost analysis itself indicates that a potential for opportunistic abuse is intrinsic to authority relations. (Dow 1987, p. 20)

Williamson denied Dow's charge, but his rejoinder was less than convincing (Williamson 1987).

Why would workers be ready to take advantage of their employers by behaving opportunistically, while employers would not engage in similar behavior? Do workers have a cruder psychology or morality than employers? But then, why would Williamson allow that firms act opportunistically with respect to other firms, but that these same firms would necessarily behave honorably toward their workers?

In a sense, Williamson may have had a point after all. In our present society, employers feel that they have a rightful claim to all the potential effort of their employees. Other than not paying workers their wages (which does occur at times) or misrepresenting work arrangements, what sort of opportunistic behavior could benefit employers vis-a-vis their workers?

Outside of violating the legal rights of their employees, employers are "justified" in demanding just about anything from their employees.

On a deeper level, some opportunistic behavior by workers might just be the expected response of people who find themselves in a competitive society where people are left to fend for themselves. If the boss refuses to pay a reasonable wage, why not attempt to correct matters by pilfering something from work. Given such attitudes, bureaucratic hierarchies might appear to be a rational defense against opportunistic behavior.

Despite the limitations of his perspective, Williamson remains an important source of understanding the deficiencies of the current economic organization. He did an excellent job of directing our attention to what he describes as opportunism. Such behavior includes "the full set of *ex ante* and *ex post* efforts to lie, cheat, steal, mislead, disguise, obfuscate, feign, distort and confuse" (Williamson 1985, note p. 51), even though, as we mentioned before, Williamson is not willing to concede that such behavior may be as common at the top of the hierarchy as at the base.

Of course, my interpretation differs from Williamson's. For him, opportunism is merely counterproductive behavior, requiring hierarchical organizations to contain it. For me, it is a symptom of a deeply flawed society in which the emphasis on self-interested, individualistic behavior smothers many, if not most, socially directed impulses. In this environment, authoritarianism is a natural outcome.

Recall my earlier discussion of David Gordon's scatter diagram, which indicated a strong relationship between the share of administrative and managerial effort and the rates of incarceration. In Gordon's perspective, an irrational lust for control makes itself felt both in the workplace and in the judicial system. My reading of Williamson suggests a parallel phenomenon. Hierarchical authority breeds opportunism and snuffs out more altruistic forms of behavior (see Frey 1997 in this regard), probably creating still more opportunism that calls for an even more powerful hierarchy.

This spiral of control leads ever more to intrusive hierarchies. I have described this process elsewhere (Perelman 1998).

Conclusion

I regret that do not have much to report on how economists have approached the subject of waste. Economists have rarely displayed any interest in this subject. While the work of McCloskey, Coase, and Williamson throws some light on the nature of waste in our economy, such results were unintentional.

Following the lead of Coase and Williamson, many economists have explored the difficulty of organizing economic activity, say, in writing effective contracts to ensure an appropriate outcome. George Akerlof's work on lemons feeds directly into this literature.

Sadly, none of this literature to my knowledge suggests the possibility of transcending these problems through a different organization of society. While even the smallest government program is sure to come under critical scrutiny, the defects of the market itself remain unexamined.

PART 3

Transforming Society

CHAPTER 9

Human Nature

Uncharted Territory

In this chapter, I will first address the question of human nature and then offer some evidence that people can indeed transcend the narrow limits that human nature supposedly imposes. Herbert Simon proposed: "Nothing is more fundamental in setting our research agenda and informing our research methods than our view of the nature of the human beings whose behavior we are studying" (Simon 1985, p. 303).

The basis of this book is my strongly held belief that everybody has the potential to be the world's best at something. The goal of society should be to assist people in discovering and then realizing that potential. Few people can take full advantage of their potential in the context of our present society. To reach this goal, as I just mentioned, we must make a thoroughgoing transformation of our society.

This transformation would not merely aim at providing people with a higher standard of living. It would entail revolutionary changes in the way that our economy organizes work. More fundamentally, it would require tapping into aspects of the human personality that are absolutely foreign to that pitiful creature—economic man.

In the wake of this transformation, our society would become unrecognizable. After all, people who have the opportunity to both find and realize their own gifts are not likely to feel the need to rob and maim their neighbors. They will not be likely to become caught up in a mad scramble to accumulate wealth and power. In such a society, people would not have much need for government regulations to prohibit people from harming either each other or their environment.

George Akerlof, together with his wife, Janet Yellen, who later became a former governor of the Federal Reserve Board and then the chief economic advisor for President Clinton, analyzed a somewhat similar situation (Akerlof and Yellen 1993). They described a three-party game among potential criminals, the law, and the community. In their game, the community decides whether or not to cooperate with the police based on their fear of retaliation, hatred of gang activities, as well as their judgement of the fairness of police. Gang members can profit from crimes so long as they elude the law. If they fail, they suffer from punishment. Finally, in their game, the government wants to minimize both crime and spending on police.

They consider three alternative outcomes. In one regime, punishment levels are severe. The community considers punishment to be unfair and refuses to cooperate with police. As a result, crime rates are high. In another regime, the government sets punishment levels low enough to be considered fair. Even so, some crime exists because consequences of punishment are not severe. In the third, norms of cooperation are so high that no crime exists. In their model, all three outcomes are possible. Once any of the three regimes comes into being, it can be stable.

Of course, in the real world, a few people will still engage in antisocial behavior. In most cases, a family or a neighborhood would be able to hold such tendencies in check. However, since the need for punishment would be so rare, society could afford humane treatment and serious efforts at rehabilitation for the few people who required incarceration.

Unlike an Orwellian society where the government imposes behavior on people who may not share the official vision of the good society, I am suggesting the possibility of a community that develops its ethical norms from the grassroots. Since people in such a community will be able to find fulfillment in their daily lives, they will be free from the anger and hostility that makes people feel the need to impose a rigid set of values on others. As a result, in such a world, minorities need not fear having to submit to the tyranny of the few.

I will put off the discussion of the possibility of such a society until the next chapter. Instead, in this chapter, I will first discuss the challenges involved in transforming society. I will also briefly discuss some of the pit-falls encountered in past efforts to remake society. Given such complica-tions, suggesting that we could somehow simply eliminate the wastes and shortfalls that we discussed in earlier chapters might seem utopian. In this sense, we might say that what has gone before promises too much.

In another more profound sense, the following chapter will suggest that our analysis has promised too little. People can and do rise far beyond what might otherwise seem to be their potential limits. Here I would ask that you recall our early discussion of the woman who lifted her car to rescue her baby.

By creating a world of what we will call "passionate labor," we can make possible a society that can offer a quality of life far beyond what we can imagine when we think of the world from the perspective of society as it now stands. I confess that I do not have a plan to get there from here, but the more of us that realize that such a potential exists, the closer we can come to meeting this potential.

The Economy of Fear

Rather than developing methods of encouraging people to discover their own potential for passionate labor, contemporary society insists that "cod-dling" the unfortunate is counterproductive. The proper remedy for those who fall by the wayside is to make them face up to the prospects of severe consequences for their failure. A long tradition grounds this "tough love" approach in pop psychology.

Writing in 1786, the Reverend Joseph Townsend, who identified him-self as a "Well Wisher to Mankind," proclaimed, "Hope and fear are the springs of industry" (Townsend 1786, p. 403). For Townsend, hope and fear operate in separate sectors of the economy. The wealthy respond to op-portunity in hopes of great gain. In contrast, Townsend insisted that fear had to be the primary motivation for the less fortunate: "The poor know little of the motives which stimulate the higher ranks to action-pride, ho-nour, and ambition. In general it is only hunger which can spur and goad them on to labour; yet our laws have said, they shall never hunger" (Townsend 1786, p. 404).

Surprisingly, Townsend's contemporary, Adam Smith, rejected such rea-soning. For Smith, the difference between the wealthy and the poor was more a matter of circumstance than a reflection of inherent differences. In Smith's words, "The difference of natural talents in different men is, in re-ality, much less than we are aware of; and the very different genius which appears to distinguish men of different professions, when grown up to maturity, is not upon many occasions so much the cause as the effect of the division of labour" (Smith 1776, I.ii.4, p. 28). Smith was even more emphatic in asserting that the way that people responded to incentives reflected their previous experience rather than some inherent psychological difference.

Ironically, Townsend is long-forgotten, while Smith remains the icon for the harsh attitude of the market economy. Notwithstanding the obligatory lip service to the wisdom of Adam Smith, the market economy still continues to follow Townsend, working through a combination of hope for the rich and fear for the poor.

Of course, Townsend had a point. Indeed, the market economy does work through a combination of hope and fear and these two factors operate differently for workers and employers. For those who can scrape together the wherewithal to begin a business, as we have seen, hope is probably excessive. We have seen how investors sink funds into new projects in hopes of capturing great rewards, often even in the face of enormous odds against success. Townsend was also correct that fear is the operative principle for workers.

Townsend was undoubtedly off base in suggesting that workers' psychology is inherently different from that of other people. As Smith had correctly noted, more often than not, psychology reflects objective conditions. The differences between workers and employers reflects differing their circumstances.

Workers have good reason not to be particularly optimistic about their prospects. Over time as our social system has calcified, fewer and fewer workers are able to rise to a position of wealth and power. For workers with limited prospects, hope is probably an irrational emotion that induces them to take foolish actions, such as purchasing a lottery ticket in the vain hope of becoming rich.

Of course, if labor becomes relatively scarce, workers could take advantage of the scarcity to ask for better wages and working conditions. However, business has been able to rig the system in such a way that ensures that workers have a more limited set of opportunities. To prevent workers from taking advantage of a temporary scarcity of labor, business calls upon the government to retard economic growth until the demand for labor declines to a satisfactory level. When unemployment becomes more common once again, the fear of losing a job increases and then the task of the government is once again to accelerate economic growth. This strategy ensures that workers enjoy a narrower set of options than investors do.

Fear, of course, can be a great motivator. In many circumstances, fear is a very useful condition. It can condition us to avoid hazardous conditions and can startle us to warn us of impending dangers.

Fear can also be counterproductive. Consider the following example. Most people can easily walk for 50 feet on a narrow plank that lies on the ground. Let that same plank span a ravine with a 1,000 foot drop and that same walk can become an exercise in stark terror. Why? Once we look down, we begin to think about the consequences of a misstep. The more we think, the more something natural—in this case, taking a step—becomes difficult. We begin to question our own abilities. Eventually, we can become consumed with doubt and even end up confirming our worst fears.

In this vein, Amitai Etzioni noted: "A large body of research shows that under stress people's decision-making becomes less rational" (Etzioni 1988, p. 73). As long as business relies on fear to motivate workers, their performance will be less than optimal.

The Economy of Hope

In many ways, the economy of fear presents more realism than the economy of hope. While nothing prevents the worst fears of people from coming to pass, the very nature of the economy ensures, with mathematical certainty, that the hopes of the majority will be dashed. Centuries ago, Adam Smith understood this equation, observing: "Wherever there is great property, there is great inequality. For one very rich man, there must be at least five hundred poor, and the affluence of the few supposes the indigence of the many" (Smith 1776, V.i.b., p. 709–710). An alert reader could well respond that Adam Smith never realized the degree of affluence that our modern technology would make possible for the large fraction of the population. Theoretically, redistribution of a small portion of the wealth of society could lift every citizen out of poverty and into relative comfort in an advanced nation, such as the United States.

Such a policy, however commendable, would not satisfy the hopes of most people. After all, success, like wealth, is relative. Think back to the discussion about positional goods. In this regard, Philip Wicksteed once observed:

> Whereas Napoleon might wish to encourage the belief that every soldier carried in his knapsack a marshal's baton, it was obviously impossible that every—as distinct from any—soldier could rise to the position of marshal. . . . For the

> existence of one marshal implies the existence of a number of soldiers who are not marshals. . . . If we cannot all be marshals, neither could we all belong to the servant-keeping class. (Wicksteed 1910, ii, p. 657)

Alas, while a lucky few may succeed in becoming a field marshall or a captain of industry, their success requires a good number of disappointed privates or proletarians.

The theory of a market economy presumes that, although the hopes of only select few will be met, the carrot of hope will drive the general populace to work hard enough and some even smart enough that society will accumulate a great capital stock that will allow people to enjoy the fruits of modern technology in a life of relative comfort—even if the hierarchical structure of society remains unchanged.

The Question of Human Nature

I have been trying to make the case that so long as society exists in its present form, people in general will be able to enjoy only a small part of the potential that our wondrous technology promises while many people will be condemned to remain within the economy of fear. Anyone who raises the question of moving toward a different organization of society, inevitably comes up against the objections of those who confidently invoke the limits imposed by what they consider to be human nature.

In fact, most economic debates ultimately turn out to revolve around questions of human nature. Later, I will discuss how people respond to war and disaster to suggest that human nature is indeed malleable, or alternatively, that such emergencies cause people to revert to a more fundamental, non-individualistic aspect of human nature. Unfortunately, virtually nothing can dislodge a strongly held preconception about human nature. Few traditional economists would find reason to disagree with the judgement of David Hume, who insisted: "It is . . . a just *political* maxim, that *every man must be supposed a knave*" (Hume 1742, p. 42). Unfortunately, this presumption is destructive.

Hume was discussing the creation of a constitution. As Bruno Frey, the Swiss economist noted in this regard, "the constitution designed for knaves tends to drive out civic virtues" (Frey 1997, p. 43). Instead, Frey proposed: "Civic virtue is bolstered if the public laws convey the notion that citizens are trusted" (Frey 1997, p. 44).

Of course, our presumption of knavery has consequences that reverberate throughout our communities. When we look around us, we see what we expect to find. To make matters worse, our experiences reinforce the worst aspects of our expectations. In an attempt to profit from sensationalism, our mass media continually spew out a torrent of reports of innumerable acts of cruelty and depravity. When we walk through the streets of our cities we see multitudes of broken people, while others are busy adorning themselves with garish displays of wealth. Our politicians behave like schoolyard bullies. We might easily be led to the conclusion that human nature condemns us to a brutal jungle where we tear flesh from each other's bodies.

The Constitution of the United States was an elaborate contrivance based on the presumption of knavery. The framers of the Constitution, largely inspired by Hume, carefully designed it to curb the presumed predatory instincts of the poor (see Wills 1982). Consider James Madison's speech of June 26, 1787 to the Constitutional Convention of the United States, where he worried about "the proportion of those who will labour under all the hardships of life, & secretly sigh for a more equal distribution of its blessings." Madison feared: "These may in time outnumber those who are placed above the feelings of indigence. According to the equal laws of suffrage, the power will slide into the hands of the former" (see United States Constitutional Convention 1787, p. 195). He and his fellow framers constructed the Constitution to limit the influence of those who are not "placed above the feelings of indigence."

Evolutionary psychologists give us further reason to despair transcending human nature, as it supposedly exists. They tell us that the human psyche has evolved over hundreds of millennia. Our recent period of presumably civilized society represents but a blink of the eye in the lifetime of humanity. As a result, our brain, which evolved to meet the needs of a typical hunting and gathering society, has not had time to evolve to suit the ideal of a modern society based on extensive cooperation. As Steven Pinker has written: "Our minds are adapted to the small foraging bands in which our family spent ninety-nine percent of its existence, not to the topsy-turvy contingencies we have created since the agricultural and industrial revolutions" (Pinker 1997, p. 207). For example, Robert Sapolsky reported on his observations of a pack of baboons living in a plush savanna park. Since this environment is so generous, the creatures need to devote only a small portion of the day to meeting their basic needs. They spend much of the rest of their time harassing each other, supposedly in an effort to gain rank within their community. Their behavior bears an embarrassing resemblance to certain characteristics of our own society (Sapolsky 1994, pp. 257–58).

This perspective, based on the immutability of human nature, immobilizes us. Left with an absence of an alternative vision, many of us stand idly by while our society rapidly destroys nature's bounty. Projecting our worst traits on others, we continue to squander massive quantities of wealth on prisons and on military technologies, which threaten to destroy ourselves along with nature. Children, through no fault of their own, grow up in situations of such squalor that they will have little chance of succeeding in life. Among those who are ostensibly successful, many still find themselves wasting their lives in meaningless work.

What is the potential for change? If our minds have not yet adapted to the traditional market society, how then can we hope to change our behavior dramatically enough to make an even more advanced society possible? We might be forgiven for accepting the word of those biologists who tell us that we are not far removed from the animal kingdom. Certainly, anyone in the public realm who dares to suggest that we exercise more compassion stands accused of irrationality or worse.

Yet, when we look in other directions we see evidence that people are capable of incredible acts of love and creativity. We might also take some comfort in the wisdom of those who lived before the ideology of market fundamentalism obscured our vision of the possible. Recall our earlier discussion of Aristotle's recognition that people's experience shaped their personality and that limited opportunity meant limited people (Aristotle 1908, Book 3, Section 5, 4th Para., p. 61). Before I build upon Aristotle's vision, I will first address the conventional conservative case for the market.

The Conservative Case for the Market

For centuries now, conservative defenders of the market have grounded their support for markets upon their understanding of human nature. In their view, people are naturally selfish, aggressive creatures. In Albert Hirschman's words:

> A feeling arose in the Renaissance and became firm conviction during the seventeenth century that moralizing philosophy and religions precept could no longer be trusted with restraining the destructive passions of men. New ways had to be found and the search for them began quite logically with a detailed and candid dissection of human nature. . . . La Rochefoucauld . . . delved into its recesses and proclaimed their "savage discoveries" with so much gusto that the dissection looks very much like an end in itself. (Hirschman 1977, p. 14)

Within this perspective, passions are irrational, dangerous and above all, threatening to society as a whole. Recall Adam Smith's description of members of the working class being driven by "passions which prompt [them] to invade property" (Smith 1776, V.i.b.2, p. 709).

Back in the eighteenth century, when the leading lights of society were engaged in the construction of markets, political constitutions, as well as their justification, the intellectual fashion was to see markets as providing an outlet for drives that would otherwise prove destructive. By channeling aggression and selfishness into profit-making activities, markets transform potentially destructive instincts into productive behavior that enriches us all (see Hirschman 1977).

Conservatives do not just see markets as a protective device to contain the passions. They also delight in their explanations of why markets are efficient. They insist that markets create incentives that call forth prodigious effort and creativity. By spurring people to exert themselves to the utmost, markets supposedly confer enormous benefits on society. According to the conservative world view, no government, however well meaning, could possibly match the efficiency of a free market.

Conservatives also insist that markets are conducive to morality. By rewarding effort and creativity, markets call forth the best in people. In contrast, they picture government programs as perversely rewarding laziness and sloth.

Well-intentioned conservatives also applaud the market as a bastion of fairness. Markets supposedly avoid the abuses of favoritism that plague all societies with strong governments. After all, markets have no reason to discriminate unfairly against one group or another. As a result, a purely market society would be without rent-seeking behavior. Finally, conservatives insist that even honest government bureaucrats often commit inexcusable errors that cost society dearly. Markets have no reason to fall into such errors.

Conservatives would have us believe that our problem is that we have refused to accept the wisdom of the marketplace. In truth, the conservative approach suffers from what seems to be a serious contradiction. Market discipline educates people in such a way that their morality improves, in the sense that they work hard for what they earn, but such improvement can only go so far. People might even improve to the extent that they will honor contracts and behave honestly, but market forces will never create incentives to develop a social conscience that will encourage people to behave unselfishly, or even morally.

No conservative, to my knowledge, has succeeded in resolving this tension between the common call for greater morality and denial that people can behave in any way except selfishly. Economists separated as much in time and ideology as John Maynard Keynes and Adam Smith, share this confusion.

For example, Nathan Rosenberg proposed that we should read Adam Smith's *Theory of Moral Sentiments* (1759) in light of his view of market morality. In Rosenberg's approving words, "Smith views a commercial society as one that, at the very least, can be relied upon to maintain stability in its stock of moral as well as physical capital" (Rosenberg 1990, p. 17).

In fact, Smith did advocate capitalism largely because he believed that capitalism would somehow promote patterns of behavior that would be superior to what the world had previously known (Perelman 1989), thereby creating a virtuous circle. For Smith, while the growth of markets improves behavior, improved behavior also contributes to the growth of markets.

Although Smith never used the term, "moral capital," Rosenberg's reading accurately reflects a deep confusion in Smith's perspective. Capital, as Smith and his fellow economists used the term, is a form of private wealth. Morality, while it can be a private matter, is not owned. Moreover, morality, diffuses throughout a society. While everybody in a society may not share identical moral values, certain moral norms become common in society.

Moral Capital and Social Capital

The notion of moral capital is very attractive. People would rather attribute their success, as well as the failures of others, to moral stature rather than to more mundane causes, such as power or luck. A French economist, Destutt de Tracy, writing in 1815, declared that "it is certain that our physical and moral faculties are alone our original riches" (Destutt de Tracy 1815, iv, pp. 99–100). David Ricardo, certainly the most influential economist of that period, concurred (Ricardo 1821, p. 285).

Nassau Senior, a staunchly conservative economist of the early nineteenth century, may have been the first economist to use the term, moral capital. Senior might not be well known today, but he was perhaps the most important public face of economics of his day. Senior occupied the first chair in political economy at Oxford University and was forever heading up economic studies for the government. According to Senior, England was successful because "the intellectual and moral capital of Great Britain far exceeds all the material capital, not only in importance, but in productiveness" (Senior 1836, p. 134).

Senior went on to contrast the poverty of Ireland with the prosperity of England. The poverty of Ireland was unrelated to the British conquest. Instead,

Ireland is physically poor because she is morally and intellectually poor, because she is morally and intellectually uneducated. And while she continues uneducated, while the ignorance and violence of her population render persons and property insecure, and prevent the accumulation and prohibit the introduction of capital, legislative measures, intended solely and directly to relieve her poverty, may not indeed be ineffectual, for they may aggravate the disease, the symptoms of which they are meant to palliate, but undoubtedly will be productive of no permanent benefit. (Senior 1836, p. 135)

At this point, we can see that for Senior, moral capital means the degree to which people are willing to submit to the rigors of the market. Even today, many people agree with Senior. For example, the social infrastructure of Robert Hall and Charles Jones, discussed in the first chapter, is quite similar to Senior's moral capital.

Similarly, Douglass North, a Nobel Prize-winning economist contends: "The inability of societies to develop effective, low-cost enforcement of contracts is the most important source of both stagnation and contemporary underdevelopment in the Third World" (North 1990, p. 54).

The logic of moral capital confirms the conventional wisdom that any economy that defies the rules of the market is bound to pay dearly for its foolishness. Consider the approach of Robert Barro, who proposes that even democratic values must give way to the dictates of the market: "It sounds nice to try to install democracy in places like Haiti and Somalia, but does it make any sense? Would an increase in political freedom tend to spur economic freedoms—specifically property rights and free markets— and thereby spur economic growth?" (Barro 1996, p. 1). Such countries would do better to turn to strong governments that promise to give free reign to capital.

Barro worries about the "growth-retarding features of democracy" because of "the tendency of majority voting to support social programs that redistribute income from rich to poor. These programs include graduated-rate income tax systems, land reforms, and welfare transfers" (Barro 1996, p. 2).

The Moral Capital of the United States of America

Barro observes that "authoritarian regimes may partially avoid these drawbacks of democracy" (Barro 1996, p. 2; see also Barro 1997, p. 119), especially if the government is favorable to the rights of capital. Indeed, we may read Barro's fear of democracy as a distant echo of James Madison, whom we quoted earlier.

Already in the eighteenth and early nineteenth centuries, many of the most famous writers of the time believed that democracy and capitalism were incompatible. Voters would obviously use their political powers to deprive property owners of the wealth.

Despite Madison's best efforts, soon after Senior wrote of the glories of moral capital, a number of government bodies in the United States failed to repay their bonds. Between 1841 and 1843, eight states and the territory of Florida defaulted on their obligations, and by the end of the decade four states and Florida had repudiated all or part of their debts (English 1996, p. 259). Barry Eichengreen, an economist from the University of California at Berkeley, reminds us that Charles Dickens's beloved *A Christmas Carol,* written during the 1840s, caught the mood of the times. Dickens had Ebenezer Scrooge awaken from a deep sleep fearing that his secure British investments have been transformed into default-prone United States securities (Eichengreen 1991, p. 149).

In the wake of the shocking epidemic of defaults, many foreign investors saw the United States as a country virtually devoid of moral capital. The same Sidney Smith, who railed against the wasteful educational system of England, caught the mood of the British in a series of famous diatribes first published in the *London Morning Post.* Dripping with moral indignation about the small part of his fortune he had invested in Pennsylvania bonds, Smith exploded:

> There really should be lunatic asylums for nations as well as individuals. . . . [America] is a nation with whom no contract can be made, because none will be kept; unstable in the very foundations of social life, deficient in the elements of good faith, men who prefer any load of infamy, however great, to any pressure of taxation however light. (Sampson 1982, p. 49; citing Pearson 1934, p. 295)

Few believed that the United States would emerge unscathed from this defiance of the rules of the market. Even though the United States Constitution precludes suits against states to enforce the payment of debts, *The London Times* predicted on December 3, 1846 that the defaulting states would have no choice but to repay their debts. The learned analysts of the paper assured their readers that the states "will deem it a not disadvantageous transaction to lay out ten or twenty millions . . . in purchasing a restoration of their forfeited respectability" (English 1996, p. 272; citing McGrane 1935, p. 166).

Alas, events proved the *The London Times* wrong on two counts. First, many of the bonds were never repaid. The Council of Foreign Bondholders, formed in 1868, still champions the cause of the holders of those bonds that remain in default to this day (Sampson 1982, p. 50). Second, despite its presumed deficiency of moral capital, the United States went on to prosper.

To begin with, the United States had the good fortune to be far enough from Britain's mighty Navy that it was able to escape the fate of a poor country today that would dare display such an insufficiency of moral capital. Besides, the British, who lost the most from the default, were reluctant to retaliate economically because cotton was crucial for the British textile industry (English 1996, p. 259). In addition, the British had few alternative outlets for their funds. As a result, they could not convincingly threaten to withhold future investments for long.

In fact, soon thereafter, following a furious boom in railroad construction in the United States, British bondholders bought up an enormous mass of paper from the railroads, undeterred by the previous deficiencies in moral capital. Alas, these British investors suffered a second round of defaults. *Bankers Magazine* reported that by 1876, about 65 percent of European holdings of American railway bonds were in default (Wilkins 1989, p. 194). Even so, British investors continued to pour money into the railroads. In 1887, *The Economist* magazine, published in London, held out a ray of hope for those who had optimistically invested in railroad shares: "As the country fills up the dividends upon American railway shares may become more stable, and these securities may come to possess something more of an investment character."

The experience of the investors in the United States railroads reminds us that while the idea of moral capital might be well and good, the hope of profits can easily quell any doubts about moral capital. While I doubt the essential role of moral capital, I remain convinced about the importance of what Jane Jacobs called "social capital" in her book, *The Death and Life of Great American Cities*. While discussing the possibility of neighborhood self-government. She observed: "If self-government in the place is to work, underlying any float of population must be a community of people who have forged neighborhood networks. These networks are a city's irreplaceable social capital" (Jacobs 1961, p. 138).

Some time later, James Coleman returned to the phrase, defining social capital as "the ability of people to work together for a common purposes in groups or organization" (Coleman 1988, p. 95). According to Robert Putnam, social capital describes "features of social organization such as networks, norms, and social trust that facilitate coordination and cooperation for mutual benefit" (Putnam 1995, p. 67). So, while Nassau Senior's moral capital refers to a willingness to submit to the individualistic mores of a market society, Jacob's social capital turns our attention back to our collective responsibilities—a concept that mainstream economics virtually rules out.

Punishment and Coercive Accumulation of Moral Capital

Although the advocates of market societies take pride in the supposed individualistic freedoms associated with the market, in reality leaders in market societies more often than not champion measures to manipulate the behavior of ordinary citizens.

So while we have little to guide us in constructing a sense of social capital, we suffer from a plethora of charlatans who clamor to lead us along the path of virtue. All too often, those who claim this mantle of moral authority attribute virtually all behavioral problems to a breakdown in discipline. In their eyes, following the logic of the economy of fear, only the threat of punishment will herd the populace into desirable behavioral patterns.

In an earlier, and perhaps more naive age, reformers believed that prisons provided a site where society could directly elevate people's morals through a form of coercive programming. For the most part, their crude efforts came to naught. Seemingly more realistic, later generations of moral reformers no longer believed that prisons could elevate morals. Instead, they put their trust in a fear of punishment so severe that it could steer the populace in the desired direction.

More often than not, prisons produce the opposite of their intended effect. While the threat of imprisonment may have a deterrent effect on those who would not have committed a crime anyway, prisons tend to harden the incarcerated.

True, in a surprising number of cases, inhuman prison conditions have led the incarcerated to look within themselves. Some took a religious path and found true spirituality. Others began to read and found a gift for literature and self-expression within themselves.

Unfortunately, society is generally reluctant to acknowledge the accomplishments of those whose education comes from harsh prison conditions rather than from elite universities. As a result, even these occasional successes of the prison system generally come to naught.

Prisons, of course, are an expensive and inefficient vehicle for promoting literary or spiritual development. Even from the standpoint of economy alone, society would fare far better by offering young people a loving and caring environment in which they could enjoy an exposure to literature, science and every other avenue through which they could develop their own gifts. Nonetheless, prisons serve as a central part of the system of social control in the United States.

People rarely discuss prisons as a means of social control. Instead, the current intellectual climate associates social control with governments that stubbornly refuse to accept the dictates of the market.

The Great Proletarian Cultural Revolution

Throughout this book, I have tried to hammer home the reality that markets do not work the way that they are supposed to function. Of all the weakness of market societies, one in particular stands out: market societies set people against each other.

Unfortunately, efforts to create a more cooperative system of social organization have floundered, mostly because the powerful governments, especially that of the United States, have taken extreme measures to snuff out those societies that have dared to experiment with other methods of organizing society. Even where such societies manage to survive, the need to protect themselves against serious threats, including military incursions or revolutions, has necessitated authoritarian measures that subvert the original intent of the experiment.

The most notorious efforts to promote social capital have appeared to be especially ham-handed. Consider the experience of the Pol Pot government in Cambodia and the Great Proletarian Cultural Revolution in China. In considering these experiments, keep in mind that both regimes suffered under the dual handicaps of extreme poverty and serious political divisions. We should examine these efforts so that we can learn what a successful transition would require.

In 1966, China launched the Great Proletarian Cultural Revolution, one of the more ambitious attempts to transcend traditional market relationships in modern times. Since then China has thoroughly repudiated this period of its history. At that time, elite intellectuals had to go to the countryside to work as poor peasants in order to learn humility and respect for the people. While the egalitarian objectives of this period were laudable, many excesses occurred during this time. Some people even lost their lives.

Nonetheless, the Great Proletarian Cultural Revolution still had a very positive side. This period gave permission for previously humble people to take part in important decisions in society for the first time.

No doubt, some of these decisions were wrong, but I suspect that the mistakes of these humble people were no more frequent or more disastrous than those of the leaders of corporations and government in our own society.

John Gurley, a distinguished economist from Stanford University, previously known for his work in conventional monetary economics and his tenure as editor of *The American Economic Review,* the flagship publication of the American Economic Association, gave an extraordinarily positive interpretation to the logic of the cultural revolution (Gurley 1970). According to Gurley, "economic development can best be promoted by breaking down specialization, by dismantling bureaucracies, and by undermining the other centralizing and divisive tendencies that give rise to experts, technicians, authorities and bureaucrats remote from or manipulating 'the masses'" (Gurley 1970, p. 310).

In contrast to the dogma of conventional economics:

> Maoists believe that while a principal aim of nations should be to raise the level of material welfare of the population, this should be done only within the context of the development of human beings, encouraging them to realize fully their manifold creative powers. And it should be done only on an egalitarian basis—that is, on the basis that development is not worth much unless everyone rises together; no one is to be left behind, either economically or culturally. (Gurley 1970, pp. 309–10)

According to the logic of the market, society should single out those with the most education, talent, and promise. By funnelling resources to the best and the brightest, the economy will grow and prosperity will trickle down to the rest. The Maoist idea was to build on the worst. Society should first concentrate of the poor and the less educated. Those who administer to the needs of the downtrodden will be improved in the process.

More important, the Cultural Revolution allowed people who had always been ignored to show finally how much they could contribute to society. For example, Joshua Horn, a British doctor, depicted the changes that occurred in his hospital at the time. He described how nurses, orderlies, patients, and even patients' friends became active in the decision-making process. Although the typical orderly had no formal medical training, she or he would spend far more time with the patient than the doctor, who might have only a few minutes to spend with the patient. As a result, the orderly might have a great deal to offer in deciding what course of treatment to follow (Horn 1971, Chapter 6).

I would like to see this element of the Cultural Revolution replicated in my own society. Doctors should have the opportunity to learn from patients and orderlies, as well as from learned journals and medical schools. Teachers should be learning from students. Workers on shop floors should be able to have a say in how to organize production.

Of course, if we were to introduce such a cultural revolution in our own society, at least as it is presently constituted, the result would be utter anarchy. We would see excesses just as oppressive as those that occurred in China. A cultural revolution is possible only with a restructuring of society so that people would respect each other, seeing that they have common interests with each other.

We would have to dismantle the gates that many affluent communities erect to shut themselves off from the rest of the world, figuratively as well as literally. People would have to share in the gains as well as in the burdens of society.

Of course, we could not dream of creating such a society overnight. Even so, now that we have seen the staggering amount of waste that exists in our own society, we can recognize that we have considerable room for error in reconstructing this new economy.

The Pol Pot Solution

One of the most notorious instances of attempting to restructure society was the sad case of Cambodia. In the wake of the devastation of the Vietnam War, the communist Khmer Rouge regime of Cambodia set out to remake society in quick order. Inevitable abuses occurred.

The Khmer Rouge policy of forcibly relocating people to the countryside appeared to be a massive abuse of human rights, but it made good sense for a nation facing immediate starvation. The Khmer Rouge, however, made a serious mistake in attempting to remake society overnight. The government set out to break down existing social values, including family ties. It treated those who resisted as enemies of society. Many transgressors, along with innumerable innocent people, paid with their lives.

We know that people are capable of incredible acts of beauty, generosity, and creativity, yet the Cambodian experience reminds us that we cannot refashion society by fiat. Whatever progress we make will be slow in coming. Society can progress only by creating positive examples and positive reinforcement for younger generations to follow.

The fascist ideology proposed a different route to the creation of a coherent society. The term, "fascism," comes from Latin word, fasces, which is a bundle of sticks bound together to an ax with a projecting blade. The fasces was a symbol of authority for ancient Roman magistrates. The external force of the binding gave the fasces far more strength than any of the individual sticks could have.

The fascists were absolutely correct in the logic of the fasces. A unified society, for better or worse, is stronger than a divided one. The problem is that the fascists depended on the external force to bind people together, just as the external binding accounted for the strength of the fasces.

The fascists also believed that government and business should work together to organize society rationally; most of society was expected to conform to the needs of this rationality. After this ideology mutated into Nazism, it took the further step of calling for the outright elimination of all those who did not fit in with the fascists' preconceived notion of an ideal citizen. Pol Pot, at least, believed that people could be redeemed—at least if they did not resist too much.

These crude efforts to remake society indicate the sort of policies that we should avoid if we are to transform society. We will turn now to examples that suggest that people are indeed capable of transcending the narrow limits of what is supposed to be human nature.

Admittedly, this process of transforming society might possibly require a temporary dictatorship, but a dictatorship in the classical sense, rather than the way the term is used today. Originally, the word, "dictatorship," came from the dictatura of the ancient Roman Republic, an institution which lasted for over three centuries. The dictatorship provided for the emergency exercise of power during brief periods to provide a temporary bulwark in defense of the republic against a foreign foe or internal subversion. The dictator was a trusted citizen. The term of office was to last for six months at most. This institution worked for the Romans until Caesar declared himself dictator in permanence (Draper 1987, p. 11).

Keep in mind that such a dictatorship was not imposed on the people from above. Nor did it involve phenomena, such as the lavish Hitler rallies, that whipped people up into an irrational frenzy. Such a dictatorship would reflect the rational choice of people to accept a temporary diminution of their personal liberty for a brief period in order that they could enjoy considerably more liberty in the very near future.

Sadly, the dictatorship of the Pol Pot era was not provisional but an absolutist attempt to impose the will of the state on the people.

Responding to Disaster

While decrees are ineffective in transforming human behavior, emergencies often bring out unexpected possibilities of transcending the selfish motives that are supposedly paramount in the human psyche. For example, in the midst of a disaster, many people display enormous heroism that would have previously seemed out of character to their friends and acquaintances. In addition, people often respond to disasters by working with both a creativity and determination that we would not expect to find in a typical work situation.

During such emergencies, most people find it perfectly natural to break away from the rule of the market. As you read my last statement, you may well be asking yourself what kind of off-the-wall person would dare to make a fool of himself writing something that most people have good reason to believe to be untrue. Before you give in to your emotions, I would like you to take a moment to think about what often happens during a time of emergency.

In the midst of natural disasters, some people are able to profit handsomely from the tragedy of others. As you might expect, normally inexpensive goods might cost a great deal in the wake of a disaster. For example, after Hurricane Hugo hit South Carolina in 1989, portable generators that normally cost a few hundred dollars sold for thousands. However, most people find the behavior of such profiteers repellant. For example, a *Newsweek* article of October 30, 1989 described the doubling of bottled water prices after the 1989 San Francisco earthquake as "mindbending audacity" (p. 10; cited in Samuels and Puro 1991, p. 62).

Indeed, at other times, firms hold prices steady in the wake of disasters, despite the profits that they could earn by charging what the market might bear (Samuels and Puro 1991, p. 62). For example, Safeway refrained from raising prices immediately after the Alaskan earthquake of 1964 and continued to do so through the month of April. It raised prices in May, only after management decided that the emergency period had passed (Dacy and Kunreuther 1969, p. 116). Truck rates were lowered, but only for those commodities that could not be conveniently shipped by boat—the competitive mode of transport (Hirshleifer 1987, p. 141).

What could cause firms to resist gouging the public? Some firms might want to maintain goodwill and avoid retribution. More important, disasters sometimes create a heightened sense of community (Samuels and Puro 1991; and Douty 1972). In other words, some people cease acting predominately as owners or guardians of capital and begin to behave as members of a community.

So, we can catch a glimpse of the potential for change during these periods of emergencies when people respond to broader social purposes than just greed. At such times, the public at large also acts in a more responsible manner. For example, after the Alaskan earthquake, consumers voluntarily limited their purchases to avoid creating shortages (Samuels and Puro 1991, p. 65).

According to the logic of laissez-faire, what appears to be price gouging is merely the price system rationing scarce goods. Within the impersonal price system, the economy gauges all actions in terms of profits and losses. No harm is intended, but the personal costs of profit-seeking behavior are ignored.

Once a sense of community takes hold, then society sheds what Samuels and Puro call "the luxury of the price system" (p. 74). Instead, people act according to the direct effect of their actions on other members of society.

Creativity and War

Earlier, we discussed how economies cast aside the market when society mobilizes for war. People too behave differently during wars. Rather than performing work in a perfunctory manner, many people redouble their efforts on the job in order to contribute to the mobilization.

I do not deny that many people are unmoved by patriotic fervor. Nor would I suggest that wartime profiteering is unknown. The point is that such individualistic motives recede during times of war, while more socially-oriented behavior becomes more common.

People even leave traces of this changed motivation in the statistical residue of the times. For example, Robert Lucas, a conservative economist who won the Nobel Prize in Economics for his work in developing techniques that cast doubt on the effectiveness of government policy, estimated the average level of economic efficiency for the United States economy by calculating the trend of the ratio of output per unit of capital between 1890 and 1954. He found that, at times, for instance during depressions, the actual output per unit of capital fell below his trend line. At other times, the actual output per unit of capital exceeded the trend line.

Lucas discovered that during the war years, 1944 through 1946, the output per unit of capital surpassed the trend line by more than 20 percent. At no time, before or after, did the United States economy match this remarkable performance (Lucas 1970, p. 154).

This achievement is extraordinary because during this period, many of the most qualified workers were in the military rather than on the shop floor. The workers who replaced them had considerably less work experience. Because of decades of discrimination, the black workers who came from the South to work in Northern factories had far less education than the workers that they replaced. Similarly, many women without much experience in working for wages effectively "manned" the assembly lines.

Conventional economic theory suggests that industrial efficiency should have suffered dire consequences from this reliance on a supposedly less qualified labor force, yet Lucas shows that nothing of the sort happened. Instead, productivity soared.

Of course, we are accustomed to expecting productivity to increase during war time. War stimulates demand, which makes the economy work more efficiently. In addition, Lucas himself attributes some of the marvelous performance of the wartime economy to the use of overtime in industry.

While increased demand and overtime may have been a factor in stimulating the economy, emotional forces were also at work. War can make people pull together. War can create a sense of urgency. At times, powerful ideals can motivate working during times of war.

In the United States, World War II called forth just such a sense of idealism. People who labored in the factories in the United States during the war often did their best in order to contribute to the struggle against fascism. Such ideals were more compelling than greed for most workers.

Casey B. Mulligan, a colleague of Lucas's at the University of Chicago, found further statistical evidence suggesting the influence of ideals (Mulligan 1998). According to economic theory, only changes in monetary incentives can change behavior. When wages fall, work effort should shrink accordingly. He found that roughly ten million more civilians were employed during the war than if employment had followed its prewar trend (Mulligan 1998, p. 1040), even though "after-tax real wages of manufacturing production workers were lower in absolute terms (and even lower relative to trend) during the war years 1942–1945 than in the few years immediately preceding and following the war" (p. 1044). Mulligan reported on his efforts to attempt to develop alternative explanations within the confines of standard economic theory to interpret this bulge in employment even though real after-tax wages were falling. In every case he failed, suggesting that patriotic idealism lay behind the rise in labor force participation. In other words, more people were working than would be expected based merely on the desire to earn more wages. Instead, people were coming into the labor force to contribute to the war effort.

Eventually, the patriotic consensus frayed around the edges. Workers became frustrated seeing their sacrifices unmatched by their employers who were enjoying unparalleled profits. As a result, toward the end of the war, strike activity began to pick up. Still, the spirit of community was sufficiently strong to produce the high levels of productivity that caught the attention of Robert Lucas.

Similarly, Israel mobilized 15 percent of its labor force for the Yom Kippur War, but the Gross National Product declined only 5 percent (Maital 1982, p. 114). The decline in production might not seem to be as impressive as the experience of the United States, but remember the United States had time to adjust to the wartime demands, while the Yom Kippur War was a brief affair.

The wartime experiences of Japan and Germany offer even more powerful illustrations of the ability of people to overcome adversity. Jack Hirshleifer, an economist from the University of California at Los Angeles, reported that ten days of bombing raids during July and August 1943 destroyed half the buildings in Hamburg. Yet, within five months the city had regained up to 80 percent of its productive capacity (Hirshleifer 1987, pp. 32–33). The United States government was interested to find out what determined the effectiveness of the Allied bombing attacks. John Kenneth Galbraith, the famous Harvard University economist, assembled a team that included some of the most prominent economists in the world. These researchers found that bombing only made the Germans more resolute. According to the findings of the survey, "the air raids of 1943–1944 . . . may have kept up the tension of national danger, and created the requisite atmosphere for sacrifice" (cited in Galbraith 1994, p. 131; see also Galbraith 1981, p. 205; and Scitovsky 1991, p. 258).

On August 6, 1945, the United States Air Force dropped an atomic bomb on Hiroshima. The next day, electric power service was restored to surviving areas. One week later, telephone service restarted (Hirshleifer 1987, p. 34).

No doubt the Germans and the Japanese, like their counterparts in the United States, worked overtime to rebuild their economic capacity, but even trebling the average work day would not have sufficed to accomplish what they did. Their success in reconstructing their economy required enormous creativity and ingenuity.

We also see a more intensive development of new technologies during periods of crisis. Ordinarily, the typical large corporation is timid about exploring new ideas, yet the same people, who typically display little creativity within the confines of the large corporations, are more inclined to promote great scientific breakthroughs under the urgency of war.

Based on his reading of Japanese history, Shigeto Tsuru, an important Japanese economist, proposed the concept of creative defeat, meaning that a horrendous defeat can unleash a torrent of energy and ingenuity. The end result can be an even greater level of economic development than would have occurred in the absence of the setback. In his words: "Japan is an example of a fantastically creative response to defeat. One recalls that Schumpeter used to puzzle the students of his 'Business Cycle' course ascribing the Japanese boom of 1924–1925 to the Great Kanto Earthquake of 1923. The defeat in the last war brought about, of course, a far greater scale of devastation in the economy of Japan, necessitating a fresh renovating start in almost every aspect" (Tsuru 1993, p. 67).

Behavioral Bandwagons

While we often speak of human nature as if it were something fixed and immutable, the response to war and other disasters reminds us that in reality people have an enormous potential for change. Jack Hirshleifer pointed out the seeming unpredictability of the response in the case of disasters (1987, p. 137). For example, he reported a substantial burst of altruistic behavior in the Halifax explosion of 1917 (Prince 1920) and in the New York power blackout of November 1965 (Rosenthal and Gelb 1965), but during the Chicago blizzard of February 1967, incidents of looting occurred.

Certainly, a mix of both selfish and generous actions occurred in all disasters. What determines whether the preponderance of people will transcend their individual self-interest and give precedence to the social good? Let me use an example to illustrate my thinking on this question.

Some mathematicians create what they call cellular automata, which are like a checker board, in which what happens in any one square can affect nearby squares. Typically the patterns that arise out of this process of interaction are unexpected, just like the patterns that develop in the wake of disasters (Albin 1998).

Psychologists know that people tend to be influenced—even caught up—by what they see around them, just like what happens in the cellular automata. If a few people begin to respond to a single socially responsible act with their own positive behavior, they can influence others, creating a wave of unselfish behavior. Similarly, people who view selfish acts are likely to respond in kind (Sunstein 1997, p. 32).

Social influences do not always bring out the best in people. Sudden shifts in human behavior can move in other, less pleasant directions, when otherwise respectable people feel free to behave in cruel and violent fashion—say during lynchings or pogroms. I suspect that quite a few participants in such violence might look back in bewilderment, or even horror, at their previous behavior, which would seem to be entirely out of character.

According to Cass Sunstein, a legal scholar concerned with understanding how societies form their conception of justice and fairness:

> current social states are often more fragile than is generally thought—as small shocks to publicly endorsed norms and roles decrease the cost of displaying deviant norms and rapidly bring about large-scale changes in publicly displayed judgements and desires. Hence societies experience norm bandwagons and norm cascades. Norm bandwagons occur when the lowered costs of expressing new norms encourages an ever-increasing number of people to reject previously popular norms, to a "tipping point" where it is adherence to the old norms that produces social disapproval. Norm cascades occur when societies are presented with rapid shifts toward new norms. (Sunstein 1997, p. 34)

Anthropologists know the difficulty of finding a fixed "human nature." Behavior varies considerably from society to society. In some, people are more competitive; in others, they are more cooperative.

We have also seen that wars and emergencies often do tip norms temporarily within a given society. Since the justification for the new norms evaporates as the war or momentary emergency passes, people tend to revert back to their old behavior, even while looking back nostalgically at those exceptional moments in their lives.

With a less temporary justification of the new norms, the altered forms of behavior can become permanent, allowing people to transcend our narrow understanding of what human nature is supposed to be. As Amartya Sen, a Nobel Prize-winning economist, once speculated: "People may be induced by social codes of behaviour to act as if they have different preferences from what they really have" (Sen 1973, p. 258). After a while, they can actually incorporate these new preferences into their own mentality.

CHAPTER 10

Toward Passionate Labor

Interductory

Recall that many people found markets attractive precisely because of the supposedly just way in which markets channel and shape human behavior. Specifically, markets supposedly served to constrain passions and reward virtues, such as diligence and prudence.

In contrast, I insist that markets tend to stifle creativity and to snuff out joy. The hard face of markets might have made some sense in the distant past when economic output sufficed to provide but a bare subsistence for the vast majority of people and affluence for only a handful. Those days are long past. Harry Johnson, a very sensible conservative, once wrote that:

> economists frequently make the mistake of thinking themselves back in classical times, when life really was a hard struggle for survival and the blunting of incentives to work was a really serious matter, because it entailed a drain on the meager resources of the responsible hardworking citizens for the benefit of social drones, or else a tax on the god-given rightful income of the propertied classes for the benefit of those who should be grateful for the opportunity to find gainful employment in their service. (Johnson 1975, p. 229)

While elites often have trouble imagining working-class people responding to any motivations outside of the economy of fear, they themselves frequently take on working-class activities voluntarily for their own enjoyment. For example, wealthy people not uncommonly do wood work, restore cars or farm as hobbies rather than occupations. Supposedly they find joy in this sort of nonmarket work.

Wealthy people also defy the logic of the profit motive in paying exorbitant prices for vineyards in the prestigious Napa Valley (Morton and

Podolny 1998). In addition, they often become owners of "opinion mag-azines, professional sport teams, films, art galleries, bars, and restaurants" in order to enjoy "some feature of the organization other than its level of profitability" (Morton and Podolny 1998).

We tend to regard vanity as the driving force behind such projects, but vanity is, at least in part, a reflection of a desire for social regard. In that sense, we need not stretch our imagination much to see how the same im-pulses that lead to these vanity investments could just as easily incline peo-ple to find gratification in working for the betterment of society as a whole rather than for purely self-interested reasons.

Nonetheless, the wealthy tend to deny such motives could affect the more brutish sensibilities of the working class, whose only possible stimu-lation is the economy of fear. While the wealthy may find joy in running a vineyard or restoring an antique car, they refuse to see that the same sort of motivation could rouse the working class from their supposed laziness more effectively than the economy of fear.

A recent *Wall Street Journal* report offers another anecdote to suggest that what seems to be purely self-interested patterns of behavior might be more malleable than most people believe. The article in question describes the experience of an upscale London restaurant, Just Around the Corner, which opened in 1986. The psychological presupposition of economics is that people calculate ways to get the best product at the lowest price. The restaurant defied this logic.

Critics described the restaurant's premise as "lunatic" and "economically irresponsible." Unlike the menus at a typical restaurant, the menus at this one do not carry prices; instead, the restaurant lets customers decide how much they should pay for their meal. But 12 years and a recession later, Just Around the Corner has developed a large and loyal following, and weekend tables have to be booked weeks in advance.

Most of the patrons are not bargain hunters either. Michael Vasos, who owns four other London eateries as well as Just Around the Corner says, "I make more money from this restaurant than from any of my other es-tablishments." Rather than underpaying, the well-heeled patrons generally overpay—on average spending 20 percent more than the price of the same meal at one of Mr. Vasos's other restaurants.

The anecdotes about vineyards and restaurants hint at the possibility that people are capable of responding to a set of incentives that would seem to violate conventional economic motivations. How can we explain in terms of conventional economics, why the customers in London would willingly pay more for a meal that they need to.

What explains why people whose time may be worth $1000 per hour may want to restore a car or to do fine wood work instead of just hiring a working-class craftperson to do the job for a modest hourly wage? By the same token, I am convinced that many poor people, perhaps with a modicum of training, are fully capable of performing work for which members of the elite presently earn huge salaries.

In addition, I am certain that by changing the context of work, most people will work with a joy and enthusiasm that will produce economic outcomes far superior than anything that we can imagine. More important, such a society would be free from most of the social ills that I previously described.

Unfortunately, while we have some scattered examples of rich people restoring cars and the like, we do not have much information about the responses of ordinary people in this regard since our society rarely offers such opportunities to less privileged people. Consequently, this part of the book will necessarily be brief, given our lack of experience in going beyond the ordinary. We can only be suggestive and anecdotal. Even so, I believe that this chapter is the most important part of the book.

Farm Work versus Gardening

In order to understand the potential for transforming the economy let me use a simple example that does not require much of a stretch of the imagination. Just think of the enormous contrast between farm work for wages and gardening as a hobby. Farm work is considered to be so abhorrent in the United States that we regularly hear that only foreign-born workers are willing to perform it. Supposedly, citizens of the United States would never be willing to subject themselves to the life of a farm worker.

While farm labor may be among the hardest, most dangerous work in our society, many people regard gardening as a pleasant diversion. While the United Farm Workers Union represents mostly downtrodden workers, a good number of wealthy people are proud affiliates of their blue-blood garden clubs. Over and above the time that they spend in their gardens, many gardeners enthusiastically devote considerable leisure time to conversing or reading in order to become better gardeners. In addition, many gardeners also willingly spend substantial sums for equipment and supplies to use in their gardens.

What, then, is the underlying difference between farm work and gardening? Farm work typically entails hard physical labor, but many gardeners also exert themselves in their gardens. The difference lies in the context of gardening. Gardeners, unlike farm workers, freely choose to be gardeners. During their free time when they work in their gardens, they want to be gardening. Nobody tells them what to do. Of course, gardeners are not entirely free to follow their whims. The rhythms of the seasons and the sudden shifts in the weather dictate some of what the gardeners do, but gardeners generally accept these demands beforehand.

As the psychologist, John Neulinger says: "Everyone knows the difference between doing something because one has to and doing something because one wants to" (Neulinger 1981, p. 15). We should also keep in mind that society respects gardeners. Our newspapers regularly print features of interest to gardeners. Some even have special sections to appeal to their affluent gardening readers. All the while, the lives of farm workers generally pass virtually unnoticed. After all, in our society, farm work is not "respectable" work in the sense that well-to-do families would not approve of their children becoming farm workers. This respect contributes to the allure of gardening.

If we paid farm workers as well as those who labor on Wall Street and accorded farm workers the sort of dignity that college professors enjoy, parents might still try to steer their children away from farm work because of the frequent exposure to potentially lethal toxins. But then, if society esteemed farm workers, farmer owners would not and could not spray them with impunity.

Gardeners engage in a modest sort of passionate labor. They tend to take pride in their gardens. They work with care and joy. They can take pleasures in their surroundings and feel a part of nature.

Farm workers take orders or, if they work by the piece, they must concentrate all their energies on picking an enormous quantity of fruits and vegetables, just to make ends meet. Recall how the short-handled hoe was designed to put a quick stop to any possible reveries about the farm workers' surroundings.

Our goal in making society work for the betterment of all people would be to convert our economy from something that resembles a nation of a few farmers working a multitude of farm workers into a new kind of economy that resembled a community of gardeners, in which workers would have good reason to attack their jobs with a sense of care, pride, joy and even exhilaration.

Hackers versus Programmers

Computer hackers offer a second metaphor for the organization of work. Every so often we read of the mysterious exploits of computer hackers. The typical hacker seems to be a very young man, more often than not, a teenager. Despite the best efforts of corporations or the government, these hackers manage to break through complex security procedures. In effect, we have the spectacle of an unsophisticated teenager defeating a substantial core of highly paid computer professionals. These hackers have even managed to break through the security walls of sensitive national security computers.

Our judicial system regards the efforts of these untrained hackers as a major threat to society. True, some hacking is destructive, but most hacking is merely an exercise in curiosity. Hackers try to meet the challenge of defeating the security measures that corporations, universities, or government agencies put in place.

To accomplish their objectives, the hackers apply a combination of co-operation and creativity. Had these same efforts been directed toward the development of better programs and programming techniques, our society's productive potential could have been advanced significantly.

Because unalienated work is so rare in our society, hacking seems to be more akin to a sport than work. The activities of the hackers reflect the more general allure of work that is self-directed. Yet these hackers put in long hours of diligent labor, driven by a challenge to succeed.

We continually hear reports about the valuable contributions of computer software to our economic and social welfare. Just think of the potential benefits that we could enjoy if society could somehow tap into the creativity and energy of the small band of hackers. Think about the possibilities of an entire computer industry that could find a way to infuse itself with a comparable spirit.

Of course, many of the true breakthroughs in the computer world came from people who initially began as hackers. As the software industry has matured, programming work has become increasingly bureaucratized. In the process, the industry smothers a good deal of the creativity of its people. If we could combine the knowledge and expertise of the trained professionals with the joy and enthusiasm of the hackers, the potential of the computer industry would advance by leaps and bounds.

In a revealing study of the clumsy means whereby business taps into workers' creativity, Tracy Kidder documented the methods by which Data General Corporation organized a team to develop a new minicomputer in his book, *The Soul of a New Machine*. Tom West, the team leader mercilessly pushed his group to finish this project at breakneck speed (Kidder 1981).

In a revealing section, we learn that West drew inspiration from a videotape in which Seymour Cray described how his little company had come to build what were generally acknowledged to be the fastest computers in the world. Cray said that he liked to hire inexperienced engineers right out of school, because they do not usually know what's supposed to be impossible (Kidder 1981, pp. 58–59).

Data General seemed to find a different message in the video. The company intentionally recruited young engineers without families who would be willing to work super-long hours (Kidder 1981, pp. 65–66). In the end, the stress proved to be too much. Most of the team burnt out, never again to participate in such a project. Not surprisingly, Data General has disappeared from the economic landscape.

To some extent, both Cray and Data General had a vague understanding of the potential of passionate labor. The willingness of young people to put in long hours of intensive work reflected an enthusiasm for participating in something important. In a relatively short time, however, they learn that their efforts do not earn them respect. Instead, they are merely raw material waiting to be transformed into corporate profits. As a result, the work turns to drudgery and the enthusiasm disappears or turns to resentment.

Enriching Work Rather than Employers

A central theme of this book is the importance of enriching work rather than employers. Once we pare away the wastes, as we have seen, our economy requires relatively little work on our part—at least so long as we make no effort to repair our environmental damage. Let me suggest one of many possibilities for enriching the workplace.

In the early days of the Industrial Revolution, employers used to tout their factories as schools. Unfortunately, the only education offered by these schools was the mind-numbing experience of unremitting discipline. Suppose, though, that the workplace could actually provide education—not just a utilitarian education to make the worker capable of turning out more profit for the employer, but a real educational opportunity in which workers could follow their own interests and inclinations.

Of course, not every worker could enjoy this opportunity. Some work by its very nature is so demanding that it consumes every bit of the workers' attention. Air traffic controllers are a case in point. Nonetheless, many workers do perform tasks that would allow them to enjoy the benefits of education while on the job. Just think of the countless hours that workers spend driving cars to commute or trucks to transport goods. Such opportunities for partial inattention open a simple way to enrich the work environment.

I would love to see our government develop the wisdom to fund the production of educational radio or cassettes so that people would have the possibility to learn while on the job, at least when work does not require people's complete attention. The cost of broadcasting educational information to these people would be trifling. Besides, the reduction in boredom while people are driving might reduce traffic accidents.

Unfortunately, even if the educational materials were available, given the existing organization of work, few employers would be enlightened enough permit their workers to enjoy the opportunity to enrich their work in this fashion, especially since a more educated workforce would present a threat to the existing authoritarian structure of work.

This idea of education on the job is not new. For example, late in the nineteenth century, Samuel Gompers, the first head of the American Federation of Labor, recalled that in his youth when he worked as a cigar maker, one of the workers would be selected to read books, such as Karl Marx's, out loud. The others would pay the readers by crediting them with making some of the cigars that they themselves produced (Gompers 1925, p. 45). This environment provided a rich education to people who would have otherwise had no opportunity for further formal education.

Later, as the cigar industry replaced skilled workers with unskilled girls, the organization of work took a giant step backwards. Gompers complained to a committee of the United States Senate that the cigar industry had instituted harsh authority relations to diminish the autonomy of the young cigar workers. They even prohibited the young girls employed as cigar strippers from conversing with each other under pain of fine or dismissal (Gompers 1883, p. 16).

Basketball Again

Earlier I alluded to the enthusiasm with which young people engage in recreational sports. Even supposedly mercenary professional athletes often continue to engage in their sports for recreation long after their professional career has ended.

In contrast to work, people become passionate about their sporting activities. Why can we not make our work as fun and challenging as athletic events? Why do we not become animated at the success of our employer the way fans do when their team excels at sporting events?

Again, people voluntarily choose to participate in sports. They decide on their own to identify with their favorite teams. Why would people be inclined to identify with their employer? They might want the employer to do well enough for them to keep their job, or better yet, well enough to give them a raise. Yet, most workers are realistic enough to realize that the interests of employers and ordinary employees conflict. In fact, employers typically profit by laying off their workers or reducing salaries and benefits. In any case, a business hires people to get as much out of them for as little as possible. As a result, many people resent their employers, and not always without good reason.

Somehow we need to find a way to restructure our society so that people can identify with what goes on at their place of employment. So long as business remains the private property of individual owners who strive to maximize profit at the expense of workers, society, and the environment, I cannot see a way that such an outcome is possible.

Admittedly, some exceptional employers do take an interest in their employees. Some even exercise a beneficent paternalism. For example, the father of my college roommate managed a factory with such beneficent paternalism. His factory could undersell comparable factories run by other major corporations. He openly invited his competitors to visit his operation to discover the reason for his success.

As he expected, his visitors never discovered his secret. All they saw was an old factory with outmoded machines. They failed to notice the human factor. This man actually had some machines designed to make the work more demanding, not easier. Why in the world would anyone want to make machines more difficult to use?

Well, let us substitute the word, "challenging." By making work more challenging, workers were less likely to become bored. Besides structuring the factory to sustain workers interest in the job, he inserted himself in his workers' lives, demanding that some take night school courses to prepare themselves for a promotion or risk being fired.

We might condemn such behavior as overly paternalistic. Today, it is probably illegal as well. Nonetheless, this factory illustrates an important point: By taking people's interests into account, substantial improvement in human achievement is possible.

This sort of top-down paternalism can take us only a small distance in capturing the potential capacity of human beings. I doubt that any of his employees ever felt the elan of a magnificent athletic maneuver while working on the job, even though they might have been more engaged than comparable workers in other factories.

Blood Money

Economists vehemently resist any suggestion that people might have the capacity to change their behavior in ways that go beyond market norms.

Consider the response to Richard Titmuss, the British sociologist whom we discussed earlier in the context of the role of equity in mobilizing a people for war. More than a quarter century ago, Titmuss wrote a book about the collection of blood for transfusions (Titmuss 1971). Although the scholarship in the book was on a high level, Titmuss's point was deceptively simple. In England, the sale of blood was prohibited. The British National Health Service ran the voluntary National Blood Transfusion Service, with a conscious goal of encouraging a clean supply of blood and promoting an ethic of voluntary donations. In contrast, the United States depended heavily on commercial sources of blood.

According to the logic of economic theory, the price paid for blood should have induced more blood donations. Instead, Titmuss suggested that the commercialization of blood donation undermined the altruistic motives to donate blood. As a result, Titmuss reported that, in the United States:

> [O]ne-third of the blood was bought and sold. Approximately, 52 per cent were "tied" by contracts of various kinds; they represented the contracted repayment in blood of blood debts, encouraged by monetary penalties; some of these donors will have themselves benefitted financially and some will have paid other donors to provide the blood. About 5 per cent were captive donors—members of the Defense Forces and prisoners. About 9 per cent approximated to the concept of the voluntary community donor who sees his donation as a free gift to strangers in society. (Titmuss 1971, p. 95)

The excesses associated with blood banks that preyed upon the poor in Haiti, Nicaragua, Belize, Mexico, Dominica, Puerto Rico, Honduras, El Salvador, and Costa Rica were even worse (Hagen 1982).

On even broader grounds, Titmuss concluded that, in comparison with the British model, the United States did not fare well:

> From our study of the private market in blood in the United States we have concluded that the commercialization of blood and donor relationships represses the expression of altruism, erodes the sense of community, lowers scientific standards, limits both personal and professional freedoms, sanctions the making of profits in hospitals and clinical laboratories, legalizes hostility between doctor and patient, subjects critical areas of medicine to the laws of the marketplace, places immense social costs on those least able to bear them—the poor, the sick and the inept—increases the danger of unethical behaviour in various sectors of medical science and practice, and results in situations in which proportionately more and more blood is supplied by the

poor, the unskilled, the unemployed, Negroes and other low income groups and categories of exploited human populations of high blood yielders. (Titmuss 1971, pp. 245–6)

Douglass Starr in his book, *Blood: An Epic History of Medicine and Commerce,* noted that at the time, "Titmuss's book hit a public nerve. . . . Soon after the Titmuss book came out in early 1971, waves of exposes appeared." In early 1972, Elliot Richardson, Secretary of the United States Department of Health, Education, and Welfare, having read the Titmuss book during his Christmas vacation, directed his staff to form a task force to look at new ways of managing the American blood supply. A couple of months later, President Nixon, declaring blood to be "a unique national resource," ordered the Department of Health, Education, and Welfare to make an intensive study of better ways to manage it" (Starr 1988, p. 288).

By 1973, the Department of Health, Education, and Welfare began a "National Blood Policy" to improve the quality of blood donated. The venture largely succeeded. Between 1971 and 1980, whole blood collected from volunteers rose by 39 percent, while that from paid donors fell by 76 percent (Frey 1997, p. 83).

Even so, problems with the profit-oriented collection of blood continued. For example, according to Victor Herbert, a professor of medicine at Mount Sinai School of Medicine in New York, and chief of the Mount Sinai Hematology and Nutrition Laboratory at the Veterans Affairs Medical Center, the blood banking industry influenced the Food and Drug Administration to discard the donated blood of Americans with iron overload, or hemochromatosis. Blood industry leaders allege (with no confirming data) that this iron-rich blood is unsafe, whereas, in fact, most Americans who need blood would benefit greatly by receiving this iron-rich blood (Herbert 1999).

Even more ominously, a recent scandal associated with the collection of plasma from prisons in Arkansas recalls the worst abuses that Titmuss reported decades ago. As a result of this tainted blood, at least, 42,000 Canadians have been infected with hepatitis C, and thousands more with the HIV virus. More than 7,000 Canadians are expected to die as a result of the blood scandal.

More than 20,000 tainted-blood victims with hepatitis C filed a class-action suit against the Canadian government, alleging that sloppy screening protocols allowed tainted blood products to make their way into Canada. In December 1998, the Canadian government established a $1.1

billion (Canadian) fund to compensate some hepatitis C victims, but advocates say the fund is insufficient (Parker 1998; Depalma 1999).

The collection and distribution of blood may be one of the more inherently communal activities in our society, in the sense that we incorporate part of other people's bodies into our own. While the legal structure in the United States allows people to treat blood as a commodity to buy and sell, thus far the purchase and sale of human organs remains illegal. Because of the ongoing shortage of human organs, a growing number of people are calling for the creation of a market for human organs.

A counter movement is also afoot. Because of the threat of blood-borne diseases, more and more people are attempting to store their own blood prior to an operation rather than using either donated blood or commercial sources—asserting a form of individualism.

Titmuss versus the Economists

Not surprisingly, Titmuss's book encountered harsh criticism from economists. You might expect an unwelcome reception by conservative economists, whose training has taught them to believe that markets are the most efficient method of managing human affairs. After all, Titmuss insisted that markets for blood are necessarily inefficient.

Yet, the most prominent reviewers of the book were not conservative, but liberal economists. They never questioned Titmuss's findings that markets could be inefficient. They did not even challenge his claims that the commercial sector was passing on tainted blood. The documentation on that scandal was too strong. No, Titmuss committed a far more serious sin in the eyes of these economists. He called into question the behavioral assumptions that form the foundation of conventional economic theory.

Economists of a liberal bent readily accept that tinkering with some markets can improve economic performance, but they also accept without question the same narrow interpretation of economic motivation that laissez-faire economists do. In fact, they may even be more protective of this element of economics lest their conservative brethren challenge their credentials by lumping the liberals together with those who fall outside the pale of conventional economics.

Rather than open themselves to ridicule by their more orthodox practitioners, liberal economists behave as if they feel the need to be more Catholic than the Pope in fending off those who would challenge the most cherished beliefs of economists. For this reason, Titmuss's harshest critics among economists came from the left/liberal spectrum of academic thought. For example, Robert Solow and Kenneth Arrow, both identified with the politics of the liberal wing of the democratic party and both winners of the Nobel prize for economics, threw down the gauntlet (Solow 1971; and Arrow 1972).

While both Solow and Arrow readily accept the existence of certain situations in which government policy should reign in market forces, they are not critical of markets as such. Instead, they fervently adhere to the basic principles of mainstream economic theory in which markets are understood to be the appropriate mechanism for guiding most social policy. Finally, both have a tendency to respond vigorously when facing a challenge to fundamental market principles.

Neither chose a technical economics journal to make his case. Arrow published his piece in *Philosophy and Public Affairs,* while Solow used the *Yale Law Journal.* After all, they could count upon virtually every professional economist to reject Titmuss's ideas. Instead, they seemed to aim for influential audiences who were literate in economics, yet vulnerable to Titmuss's seductive approach.

Solow acknowledged that Titmuss wrote an "extraordinarily interesting book," which "probably is . . . a devastating and unanswerable indictment of the American system" of blood collection. However, Solow insisted on limiting the scope of Titmuss's critique: "From this secure beachhead, Professor Titmuss launches a broad attack on the market as a mechanism for the mobilization and allocation of scarce resources, and, even more generally, on the whole economizing mode of thought" (Solow 1971, p. 1696).

Solow strongly condemned Titmuss's idea that markets are antithetical to altruism (Solow 1971, pp. 1703–1704). He admitted: "In their enthusiasm for market allocations, many economists seem to drift into assertions that go beyond what economic analysis will support" (Solow 1971, p. 1707).

Solow charged that Titmuss also went too far in suggesting that defects in the market for blood are evidence of deeper questions about economic theory. He accepted that "some of Titmuss's complaints about the enthusiasts of the market are valid. But they are valid complaints about invalid overextensions of economic reasoning, not about economic reasoning itself" (Solow 1971, p. 1709).

While Solow began his review on a positive note, he ended with a nasty ad hominem remark: "There is a slight, rather typically Fabian, authoritarian streak in Titmuss; he seems to believe that ordinary people ought to be happy to have many decisions made for them by professional experts who will, fortunately, often turn out to be moderately well-born Englishmen"

(Solow 1971, pp. 1711). In short, you can call for state action to modify defects in the market, but don't dare call basic economic theory into question.

Arrow's review was less intemperate. He framed his critique in terms of what he considered to be common sense:

> Economists typically take for granted that since the creation of a market increases the individual's area of choice, it therefore leads to higher benefits. Thus, if to a voluntary blood donor system we add the possibility of selling blood, we have only expanded the individual's range of alternatives. (Arrow 1972, pp. 349–50)

Then he asked incredulously, "Why should it be that the creation of a market for blood would decrease the altruism embodied in giving blood?" (Arrow 1972, p. 351). Earlier, I quoted Arrow as writing, "An economist by training thinks of himself as the guardian of rationality, the ascriber of rationality to others, and the prescriber of rationality to the social world" (Arrow 1974, p. 16). This exchange with Titmuss casts some light on what Arrow might have meant.

Titmuss described a different sort of rationality than Arrow proposed. His implicit answer to Arrow's question ran through his book: once people associated blood donations with the behavior of winos and other less elite sort of folk, the donation of blood seemed to be a less noble gesture. In other words, the institutional framework changed the underlying behavior of potential blood donors.

Of course, the commercial sector still exists, yet many people still do donate blood, managing to separate the commercial blood business from their own altruistic acts. To the extent they make this separation, Arrow's beloved economic theory escapes unscathed. Titmuss's data, however, indicated that the commercialization of the blood system had indeed diminished altruism. Later events also seemed to bear out Titmuss's analysis.

The relative decline of commercial blood collection suggests that Titmuss had a better handle on the situation than the Nobel laureates. With the decline in commercial blood collection, giving blood once again became associated with altruism.

Titmuss had one advantage over the economists, whose training teaches them to presume that human behavior is fixed and unchanging; and that people only respond to purely economic incentives. Titmuss went further than merely challenging the behavioral assumptions of economic theory. He even proposed that nonmarket institutions, such as the organization of voluntary blood donations, can elevate people's motives to become even more distant from the narrow behavior that economists assume. He speculated:

> Men are not born to give; as newcomers they face none of the dilemmas of altruism, and self-love. How can they and how do they learn to give—and to give to unnamed strangers irrespective of race, religion or colour—not in circumstances of shared misery but in societies continually multiplying new desires and syndicalist [sic] private wants concerned with property, status and power? (Titmuss 1971, p. 12)

This potential to transcend the circumscribed bounds of individualistic rationality is the key to bringing the economy, as well as society in general, up to its true potential.

Individualism

The idea of transcending individualistic rationality usually suggests an image of an autocratic organization coercing hapless individuals. For example, consider the statement of the usually progressive J. M. Clark:

> Individualism does not assume that the individual is perfectly intelligent, for a nation of such individuals could organize a socialistic state successfully and avoid the admitted wastes of individualism, but it assumes that he has enough intelligence to look after his own interests in direct exchanges better than some outsider can do it for him, and also to learn by his mistakes, thus converting them into essential incidents of the only life worth living: the life of a being who makes decisions and takes consequences. (Clark 1939, p. 151)

Clark made the most enlightened case possible for individualism. Unlike most defenders of individualism, he explicitly takes the learning process into account. Even so, Clark's individualism still comes up short. Our actions do not take place in isolation. They affect others, even society as a whole. Consequently, the learning process requires a structure in order to create the appropriate feedback. Even more important, social forces affect the formation of our behavior patterns. We have already spoken of the role of behavioral bandwagons.

To transcend individualism does not require a totalitarian system that imposes discipline on the rest of the people. The alternative is a cooperative structure, in which people organize together in order to decide what is best for the community without impinging on individual rights.

Admittedly, this ideal has the potential to spill over into the society in which the majority demands conformity from everybody else. I would argue that this undesirable form of group pressure is more likely to come in situations in which people are insecure. When people feel more confident, they feel less threatened by those who are different. In such a society, people are more likely to recognize that the diversity of experiences and lifestyles enhances learning and makes possible a far better society for all.

Finally, we should take note of the false impression that individualism, as we know it in our society, is tantamount to freedom. In truth, our notion of an individualistic society entails an enormous amount of coercion. The supposedly free individual must conform to the dictates of the market, as well as the whims of the employer or the agent of the employer.

Nowhere is the sort of freedom that I envision that would allow work and play to merge more important than in the realm of science.

Science and Playful Labor

As the world economy continues to supply a growing population with (at least for many people) a higher standard of living in the face of a shrinking resources base, we have become increasingly dependent on science to devise improved methods just to maintain our present material conditions. We have already seen that we have done a poor job of putting our scientific talent to use, squandering many of our highly-trained scientists in work that does not take full advantage of their training or exploiting them in low-wage postdoctoral positions.

In fact, the improvement of our way of life requires that we train far more scientists than we presently do. We can do with fewer physicists devising financial strategies for speculators and fewer biologists driving taxi cabs, assuming that their scientific efforts could improve the quality of life. To facilitate the scientific process, we need to do more than just train people; we need to supply them with an environment in which they can be free (within reason) to explore the possibilities that arouse their curiosity.

I do not mean to suggest that science should be the narrow purview of a select society of highly trained individuals. I am sympathetic to the ideals (as opposed to the practice) of the cultural revolution in which society learns to tap into the creativity of all people. Let me drop that idea now and return to how market forces treat science.

We have already seen that the market has not been able to absorb as many scientists as our universities are training; however, marketplace discipline has little place in the scientific world. Science is an arena in which playful labor is probably the most effective labor (see Huizinga, p. 203).

Serendipitous discoveries from seemingly frivolous interests are the order of the day.

After all, what draws people to science in the first place? Few, if any scientists have followed the lure of monetary gain in choosing a scientific career. Most accounts that I have read describe the future scientist as having been captivated by a mystery that he or she wanted to understand at an early age. As children, these future scientists seemed to have recognized what administrators today do not—that scientific exploration can and should be a joyful experience; that the thrill of discovery is an essential element of the scientific process.

Maybe this situation will change in the future, now that some superstar scientists are beginning to command substantial salaries or to earn fortunes by exploiting the monetary value of their discoveries. Perhaps, some time in the future young people will be drawn to science rather than finance as the most obvious path to wealth. Even so, I cannot believe that pure greed will be ever be nearly as effective in promoting good science as the playful, but passionate labor of serious scientists.

Passionate Labor

Throughout the book, I have alluded to Charles Fourier's concept of passionate labor. Fourier was a brilliant, but quirky utopian philosopher of the early nineteenth century. Fourier was not an economist. In fact, his orientation would be absolutely foreign to the mainstream of modern economics. Instead, Fourier was a keen critic of society. As Frederick Engels wrote of him: "Fourier is not only a critic, his imperturbably serene nature makes him a satirist, and assuredly one of the greatest satirists of all time" (Engels 1894, p. 308).

Fourier realized that the typical worker of his day, no less than those of our own, was mired in an alienating environment that did nothing to inspire great achievement. He thought that the purpose of life, including work, was to promote happiness, including the happiness of the worker. In this spirit, he wanted to transform society from top to bottom in order that work, as well as life in general for that matter, could become an experience of joy. According to Fourier:

> In order to attain happiness, it is necessary to introduce it into the labours which engage the greater part of our lives. Life is a long torment to one who pursues occupations without attraction. Morality teaches us to love work: let it know, then, how to render work lovable, and, first of all, let it introduce

luxury into husbandry and the workshop. If the arrangements are poor, re-
pulsive, how [sic] arouse industrial attraction? (Fourier 1901, pp. 165–6)

Elsewhere, he wrote: "Civilized pleasures are always associated with unproductive activities, but in the [what he called] societary state varied work
will become a source of varied pleasures" (Fourier 1971, p. 276). If society
had followed Fourier's lead and made work less repugnant, economic productivity would exceed today's performance by leaps and bounds, even
though heightened productivity would be an incidental by-product of
providing workers an opportunity to enjoy their work.

In contrast, the happiness of workers does not enter into modern accounts of the economy. Instead, contemporary economists measure the
success of an economy solely in terms of the production of marketable
goods and services. Satisfaction on the job is relevant only insofar as it
serves to spur productivity, reversing the order of importance according to
Fourier.

Fourier wanted to go well beyond satisfaction. He believed that the
drudgery of work should give way to the possibility of passionate labor.
Here are some examples of his breathless presentation of passionate
labor:

> An uprising breaks out, barricades are needed, it is difficult labor; and how
> are the workers paid? By blows from rifles: no matter: the intrigue, the at-
> traction are there, the barricades rise as if by magic; after the victory one
> says: how was that enormous labor accomplished in three days by workers
> who are shot at instead of paid, and who worked by night without anything
> to eat? (Fourier 1835, vol. 8, p. 400)

He continued:

> Thus, the man of cold blood, returned to his reason, is not able to conceive
> of the labor he did in the fire of attraction, he cannot believe it himself. The
> French soldiers were not able to repeat, in the form of an exhibition, the as-
> sault of Mahon which they had executed the evening before under the fire
> and the rocks of the enemy. One writes from Liege, after the submersion of
> 80 miners in the Beaujon-Goffin pit, "what they did in 4 days was unbe-
> lievable, paid workers would not have been able to do it in 15 days." And
> the workers, far from being moved by the appetite for gain, were indignant,
> felt outraged when one spoke to them of payment; the sacred fire, attraction
> was there, charity, amity, the honor of saving comrades, close to death.
> (Fourier 1835, vol. 8, p. 401)

Fourier returned to the subject of the trapped miners a few years later:

> This type of bond is excessively rare in the civilized world. It does not show up unless fortuitously and in glimpse; but in its short appearances, it raises men to a state which could be called ultra human perfection: it transforms them into demi-gods in whom all wonders of virtue and industry become possible. . . .
>
> Their companions, electrified by friendship, worked with a supernatural ardor and were offended by the offer of pecuniary compensation. They accomplished in order to free their buried companions, wonders of industry about which the reports said, "What was done in 4 days was incredible. People in the mine business assured others that, if wages were involved, this amount of work could not have been obtained in 20 days."
>
> What is this impulse that suddenly gives birth to virtues, wonders of industry united with disinterestedness? It is . . . a vehement passion, a fiery virtue, it is only a sacred fire. . . . These workers, coming from other pits, did not know individually those (trapped) in the pit of Beaujonc. There was nothing personal in this devotion; it was collective philanthropic affection and not individual. (Fourier 1838, vol. 3, p. 373)

A few decades later, the author of a treatise on the exploitation of coal mines, published in Liege, offered a different evaluation of the coal miners when working under more normal conditions. He complained that "workers, having no interest to work actively, mostly slacken the sum of their efforts as soon as supervision ceases" (Ponson 1854, p. 120; cited in Hobsbawm 1986, p. 353).

So, the disinterested miners became passionate miners and then disinterested once again, depending on circumstances. We see people rise above the level of disinterest, time and time again: the mother lifting the car to rescue her child, the self-sacrificing hero on the battlefield, or the amateur athlete pushing himself or herself to the utmost.

To my knowledge, Fourier was the first person to explore the possibility of basing an entire society on a regime of passionate labor. Admittedly, Fourier's vision of tapping into passions went far beyond my superficial discussion of his ideas.

The problems that we face today are more threatening than those that confronted people in Fourier's day. In this sense, what was utopian then may be practical, if not necessary, today.

Perhaps Fourier's vision might not be that difficult to achieve after all. For example, Donald Hebb described a celebrated Canadian experiment in which 600-odd pupils of a Montreal primary school were suddenly told that they no longer had to attend classes unless they wanted to, and that punishment for misbehavior would henceforth consist in being sent to the playground to play. All of the children dashed out of the school, but within two days they were all back in class, on a somewhat less regular schedule than before, but doing no less, and sometimes better work (Scitovsky 1976, p. 91; citing Hebb 1930; and 1955).

CHAPTER 11

Concluding Remarks

The Challenge of the Present

The central thesis of the first part of this book contends that the technological potential of the economy that exists today in advanced market economies remains unrealized. Instead, market forces either leave too many people behind with a lack of training and/or employment, while they channel productive energies into unproductive activities. This book reviewed a small sample of the wastes and missed opportunities that pervade present day market society.

The second half of the book points in the direction of Fourier's vision of passionate labor. It suggests that workers who are unemployed, uninspired, and/or resentful of their situation will contribute only a small portion of their potential to socially beneficial outcomes.

Some of Fourier's long forgotten notions about sex, as well as some of the specifics of his detailed plan of social organization might seem ridiculous today, but his vision of passionate labor is more relevant than ever. In fact, I suspect that future generations looking back at our current system will see Fourier as being far more practical than anything our present situation represents. After all, what could be more ridiculous than to consign the working day to be a veritable black hole of pleasure from which only tired workers and marketable commodities emerge?

Whatever pleasures the world has to offer will supposedly come in the form of consumption goods purchased on the market. Why in the world should we point with pride to the outpouring of commodities that flow from the modern workplace while forgetting the workers who sacrifice a good part of their lives to producing these commodities?

Despite all the rhapsodic talk of the information age (see Perelman 1998), within the prevailing organization of the economy, progress still relies on the outmoded process of crudely harnessing more and more of the worlds' population to the heavy yoke of the workplace, without any regard whatsoever for the workers themselves. In this sense, we have not progressed much since the time when the Egyptians built the pyramids.

While we can lament the fate of those who labor in the toxic stews of our manufacturing processes or even those who find themselves confined to the numbing boredom of meaningless activity in a comfortable office, we should not see these people as mere victims of a heartless system. A victim's loss is an individual loss. True, this loss may also affect a small circle of family or friends. The issue here is much larger than victimization. Victims earn our pity, but rarely prod us to seriously think about larger issues.

Following the insights of Fourier, we realize that we are all the losers in this archaic system. Had our society given the apparent victims the opportunity to excel in some form of passionate labor, everybody's lives could be enriched in ways that we cannot even imagine today.

How can we afford to allow our bureaucracies and hierarchies to continue to snuff out people's creativity? Should we not demand a radically new type of society that would not just permit, but actively promote, passionate labor?

Of course, morality and values cannot be legislated. Nor can we "reform" society by passing a few "humane" laws. Even less can we call upon the supposed generosity of well-intentioned business people to improve our system of organizing humanity. Least of all can we expect much from calling for government reforms from above. As Karl Marx, who thought long and hard about such matters, wrote in one of his early works:

> If we imagine that decrees are all that is needed to get away from competition, we shall never get away from it. And if we go so far as to propose to abolish competition while retaining wages, we shall be proposing nonsense by royal decree. But nations do not proceed by royal decree. Before framing such ordinances, they must at least have changed from top to bottom the conditions of their industrial and political existence, and consequently their whole manner of being. (Marx 1847, p. 147)

The Challenge of the Future

The hardest part of creating a regime of passionate labor will be to develop habits of tolerance and respect. This transformation will take considerable time. After all, our prejudices have been nurtured for many centuries. They will not disappear over night.

Once we learn to expect the best from others, to encourage their creativity, we will begin to discover the joys of passionate labor. The rest will be easy.

No sane person can think that this transformation of human behavior will be a simple matter. Even Karl Marx, who was prone to see revolutionary potential around the corner, including circumstances where none existed, warned in the strongest possible tones that the real struggle would begin after we freed ourselves from the fetters of market society.

Similarly, John Maynard Keynes, thinking that capitalism would have to undergo dramatic changes in the near future, commented, "I think with dread of the readjustment of the habits and instincts of the ordinary man, bred into him for countless generations, which he may be asked to discard in a few decades" (Keynes 1930, p. 327).

Like Marx and Keynes, I do not minimize the difficulty in creating an environment in which people could all live together harmoniously and with minimal problems. Even though a new psychology might be necessary for social survival, market society conditions people into selfishness. In fact, selfishness may well be is a prerequisite for economic survival in a market society. People do not abandon their habits overnight.

Recall the fate of the biblical Jews, who had to wander for 40 years in the desert until one generation, which was not yet ready for the Promised Land gave way to a new generation unburdened by the old values. Not even Marx believed that a complete transformation would come quickly. Similarly, a long period of turmoil will have to persist before people will be able to go a long ways toward shedding their selfish ways. According to Marx and Engels, writing in 1846, remarked "only in a revolution [can a people] succeed in ridding itself of all the muck of ages and become fitted to found society anew" (Marx and Engels 1846, p. 53). Four years later, Marx predicted: "You will have to go through 15, 20, 50 years of civil wars and national struggles not only to bring about a change in society but also to change yourselves" (Marx 1979, p. 403).

Toward Capturing the Potential of Passionate Labor

What could we do in the short run to move our society in the direction of a regime of passionate labor? The first order of business would be to rid society of all of those forces that restrict people from being able to discover their human potential. The next step would be to provide institutions to facilitate communication and cooperation. We need to demand that the government truly represent the interests of its citizens would above all serve as facilitator.

To close on an optimistic note, recall the burst of productivity during World War II. Certainly, every single individual did not suddenly become socially conscious. However, enough people did become socially conscious or patriotic enough that wartime productivity soared.

Hopefully, some sort of bandwagon effect can begin to take place rather quickly. Undoubtedly, some people will prefer the the status quo. Their resistance might well offset the potential gains from those who would be enthusiastic about the transformation. If such people gained the upper hand, they might discredit the transformation.

This initial period will be crucial. If society can withstand the early turmoil, the transformation will have a good chance of succeeding. Difficult as the transformation of society might seem—whether it proceeds along the rather modest course I am suggesting here or the revolutionary path that Marx foresaw—society has no choice but to begin the process as soon as possible. Otherwise, the wastes that pervade our economy will continue to swell. People will continue to suffer degradation both in the workplace and in the marketplace, while their latent talents wither undiscovered. All the while, environmental destruction will accelerate. To be blunt, our current path leads toward more wars, more devastation, and, perhaps, even our ultimate extinction.

So long as we continue to allow a few individuals and corporations to continue to monopolize the world's productive capacity in order to increase their own private wealth, society will never realize even a smidgen of the potential of passionate labor. The alternative path points to an untapped reservoir of productivity.

In our contemporary world, people often refer to the importance of freedom. What kind of freedom confines the vast majority of people to boring and even dangerous work rather than developing their potential? What kind of freedom encourages people to fulfill themselves in mindless consumption? Marx once noted: "The true realm of freedom . . . [is] the development of human powers as an end in itself" (Marx 1981; 3, p. 959).

More than a century and a half ago the British historian, Thomas Babington Macaulay, wrote an essay on Lord Clive. Looking back on an indiscretion in Clive's life, Macaulay concluded that "he committed, not merely a crime, but a blunder" (Macaulay 1840, p. 341). Similarly, recalling the way our economy treats its workers, some future commentator will certainly be justified in making an analogous judgement—that the current method of organizing our economy is "not merely a crime, but a blunder."

References

Aglietta, Michael. 1979. *A Theory of Capitalist Exploitation: The U.S. Experience,* trans. David Fernbach (London: New Left Books).

Akerlof, George. 1970. "The Market for 'Lemons': Asymmetrical Information and Market Behavior." *Quarterly Journal of Economics,* Vol. 83, No. 3 (August): pp. 488–500.

Akerlof, George A. and Janet Yellen. 1993. "Gang Behavior, Law Enforcement, and Community Values," in Henry J. Aaron, Thomas E. Mann, and Timothy Taylor, eds. *Values and Public Policy* (Washington, D.C.: Brookings Foundation): pp. 173–209.

Albin, Peter S. 1998. *Barriers and Bounds to Rationality: Essays on Economic Complexity and Dynamics in Interactive Systems* (Princeton: Princeton University Press).

Alesina, Alberto and Roberto Perotti. 1996. "Income Distribution, Political Instability and Investment." *European Economic Review,* Vol. 40, No. 6 (June): pp. 1203–1229.

Alesina, Alberto and Dani Rodrik. 1994. "Distributive Politics and Economic Growth." *Quarterly Journal of Economics,* Vol. 109, No. 2 (May): pp. 465–90.

Alford, L. P. 1929. "Technical Change in Manufacturing Industries," in Committee on Recent Economic Changes. 1929. *Recent Economic Changes in the United States: Report of the Committee on Recent Economic Changes of the President's Conference on Unemployment* (New York: McGraw-Hill): vol. i, pp. 96–166.

American Engineering Council. 1921. *Waste in Industry by the Committee on Elimination of Waste in Industry of the Federated American Engineering Societies.* 1st ed. (Washington, D.C.: Federated American Engineering Societies; NY: McGraw-Hill).

Anderson, Simon P., Jacob K. Goeree, and Charles A. Holt. 1998. "Rent Seeking with Bounded Rationality: An Analysis of the All-Pay Auction." *Journal of Political Economy,* Vol. 106, No. 4 (August): pp. 828–52.

Annable, James. 1988. "Another Auctioneer is Missing." *Journal of Macroeconomics,* Vol. 10, No. 1 (Winter) pp. 1–26.

Anon. 1988. "Airline Backfire: Texas Air Triggered Investigation of Itself with Shuttle Gambit." *Wall Street Journal* (April 15): pp. 1 and 16.

Anon. 1992. "Let Them Eat Pollution." *The Economist* (February 8–14): p. 66.

Arendt, Hannah. 1966. *The Origins of Totalitarianism* (New York: Harcourt, Brace & World).

Aristotle. 1908. *Nicomachean Ethics,* tr. W. D. Ross (Oxford: Clarendon Press).

_____. 1988. *The Politics,* ed. Stephen Everson (Cambridge: Cambridge University Press).

Arrow, Kenneth J. 1971. "Political and Economic Evaluation of Social Effects and Externalities." in M.D. Intriligator, ed. *Frontiers of Quantitative Economics* (Amsterdam: North-Holland): pp. 2–23.

_____. 1972. "Gifts and Exchanges." *Philosophy and Public Affairs,* Vol. 1, No. 4 (Summer): pp. 343–62.

_____. 1974. *The Limits of Organization* (New York: W.W. Norton).

Audretsch, David B. 1995. *Innovation and Industry Evolution* (Cambridge: MIT Press).

Audretsch, David B. and Zoltan J. Acs. 1994. "Entrepreneurial Activity, Innovation, and Macroeconomic Fluctuations." in Yuichi Shionoya and Mark Perlman, eds. *Innovation in Technology, Industries, and Institutions: Studies in Schumpeterian Perspectives* (Ann Arbor: University of Michigan): pp. 173–83.

Baldwin, James and Paul K. Goreski. 1991. "Firm Entry and Exit in the Canadian Manufacturing Sector." *Canadian Journal of Economics,* Vol. 24, No. 2 (May): pp. 300–23.

Banfield, Edward C. 1958. *The Moral Basis of a Backward Society* (New York: The Free Press).

Barber, William J. 1985. *From New Era to New Deal: Herbert Hoover, the Economists, and American Economic Policy, 1921–1933* (Cambridge: Cambridge University Press).

Barr, Stephen. 1999. "Health Care Industry Has a Lot of Work to Do." *Washington Post* (25 March): p. A35.

Barro, Robert J. 1996. *Getting It Right: Markets and Choices in a Free Society* (Cambridge: MIT Press).

_____. 1997. *Determinants of Economic Growth: A Cross-Country Empirical Study* (Cambridge: MIT Press).

Baum, Dan. 1996. *Smoke and Mirrors: The War on Drugs and the Politics of Failure* (Boston: Little, Brown).

Beesley, M. E. and R. T. Hamilton. 1984. "Small Firms' Seedbed Role and the Concept of Turbulence." *Journal of Industrial Economics,* Vol. 33, No. 2 (December): pp. 217–31.

Belcher, D. 1962. *Wage and Salary Administration* (Englewood Cliffs, NJ: Prentice Hall).

Berdahl, Robert M. 1999. "The Public University in the Twenty-First Century." Address to National Press Club (Washington, D.C. 2 June): http://www.chance.berkeley.edu/cio/chancellor/sp/press_club_address.htm

Birdsall, Nancy, D. Ross, and Richard Sabot. 1995. "Inequality and Growth Reconsidered: Lessons from East Asia." *World Bank Economic Review,* Vol. 9, No 3 (September): pp. 477–508.

Blumenstyk, Goldie. 1998. "Berkeley Pact With a Swiss Company Takes Technology Transfer to a New Level." *Chronicle of Higher Education* (December 11): p. A 56.

Bok, Derek. 1993. *The Cost of Talent: How Executives and Professionals Are Paid and How It Affects America* (New York: Free Press).

Bonzon, Thierry. 1997. "Coal and the Metropolis," in Jay Winter and Jean-Louis Robert, eds. *Capital Cities at War: London, Paris, Berlin, 1914–1919* (Cambridge and New York: Cambridge University Press): pp. 286–302.

Bonzon, Thierry and Belinda Davis. 1997. "Feeding the Cities." in Jay Winter and Jean-Louis Robert, eds. *Capital Cities at War: London, Paris, Berlin, 1914–1919* (Cambridge and New York: Cambridge University Press): pp. 305–41.

Boswell, James. 1934. *Life of Johnson,* 6 vols. G.B. Hill, ed. (Oxford: Clarendon Press).

Bourdieu, Pierre. 1984. *Distinction: A Social Critique of the Judgement of Taste* (Cambridge: Harvard University Press).

Bowles, Samuel and Richard Edwards. 1985. *Understanding Capitalism: Competition, Command, and Change in the U.S. Economy* (New York: Harper & Row).

Bowles, Samuel and Herbert Gintis. 1995. "Escaping the Efficiency Equity Trade-off: Productivity-Enhancing Asset Redistributions," in Gerald A. Epstein and Herbert M. Gintis, eds. *Macroeconomic Policy after the Conservative Era: Studies in Investment, Saving and Finance* (New York: Cambridge University Press): pp. 408–40.

Brecher, Jeremy. 1988. "Upstairs, Downstairs: Class Conflict in an Employee Owned Factory." *Zeta,* Vol. 1, No. 2 (February) pp. 68–74.

Brown, Eleanor, Richard Spiro, and Diane Keenan. 1991. "Wage and Nonwage Discrimination in Professional Basketball: Do Fans Affect It?" *American Journal of Economics and Sociology,* Vol. 50, No. 3 (July): pp. 333–45.

Buchanan, James M., Robert D. Tollison, and Gordon Tullock. 1980. "Preface." James M. Buchanan, Robert D. Tollison, and Gordon Tullock, eds. *Toward a Theory of the Rent-Seeking Society* (College Station: Texas A & M University).

Burawoy, Michael. 1979. *Manufacturing Consent* (Chicago: University of Chicago Press).

Butterfield, Fox. 1995. "More Blacks in Their 20's Have Trouble With the Law." *New York Times* (October 5): p. A18.

Caves, Richard E. 1977. *American Industry: Structure, Conduct, Performance,* 4th ed. (Englewood Cliffs, NJ: Prentice-Hall).

____. 1980. "The Structure of Industry." in Martin Feldstein, ed., *The American Economy in Transition: A Sixtieth Anniversary Conference* (Chicago: University of Chicago Press): pp. 501–44.

Clark, Gregory. 1984. "Authority and Efficiency: The Labor Market and the Managerial Revolution of the Late Nineteenth Century." *Journal of Economic History,* Vol. 44, No. 4 (December): pp. 1069–83.

Clark, John Maurice. 1917. "The Basis of War-Time Collectivism." *American Economic Review,* Vol. 7, No. 4 (December): pp. 772–90.

____. 1939. *Social Control of Business,* 2d ed. (New York: Kelley, 1969).

____. 1942. "The Theoretical Issues." *American Economic Review,* Vol. 32, No. 1 (Supplement): I: Part 2 (March): pp. 1–12.

Clark, Victor Selden. 1929. *History of Manufactures in the United States*, 3 vols. (New York: McGraw-Hill).

Clarke, G. R. G. 1995. "More Evidence on Income Distribution and Growth." *Journal of Development Economics*, Vol. 47, No. 2 (August): pp. 403–27.

Coase, Ronald. 1937. "The Nature of the Firm." *Economica*, Vol. 4, pp. 386–405; reprinted in Louis Putterman, ed. *The Economic Nature of the Firm: A Reader* (Cambridge: Cambridge University Press, 1986): pp. 72–85.

Cohen, Mark A. 1989. "Corporate Crime and Punishment: A Study of Social Foreign and Sentencing Practice in the Federal Courts, 1984–1987." *American Criminal Law Review*, Vol. 26, No. 3 (Winter): pp. 605–60.

Coleman, James S. 1988. "Social Capital in the Creation of Human Capital." *American Journal of Sociology*, Vol. 94, pp. S95-S120.

Commission on Professionals in Science and Technology. 1998. "Salary and Employment Survey: Employment of Recent Doctoral Graduates in Science and Engineering." http://www.nextwave.org/survey/.

Cowling, Keith. 1982. *Monopoly Capitalism* (New York: John Wiley).

Cutler, David M. and Lawrence H. Summers. 1988. "The Costs of Conflict Resolution and Financial Distress: Evidence from the Texaco-Penzoil Litigation." *Rand Journal of Economics*, Vol. 19, No. 2 (Summer): pp. 157–72.

Dacy, Douglas C. and Howard Kunreuther. 1969. *The Economics of Natural Disasters: Implications for Federal Policy* (New York: Free Press).

Depalma, Anthony. 1999. "Suit Says Canada Imported Tainted Blood from U.S. Inmates." *New York Times* (January 29).

Destutt de Tracy, Antoine Louis Claude. 1815. *Elemens d'Idelogie* (Paris: Courcier).

Diamond, Jared. 1997. *Guns, Germs, and Steel: The Fates of Human Societies* (New York: W.W. Norton).

Direct Marketing Association. 1999. "Total DM Employment by Medium and Market." <www.the-dma.org>

Doeringer, Peter and Michael Piore. 1971. *Internal Labor Markets and Manpower Analysis* (Boston: D. C. Heath).

Dorfman, Joseph. 1940. *Thorstein Veblen and His America* (New York: Viking).

Dorman, Peter. 1996. *Markets and Mortality: Economics, Dangerous Work, and the Value of Human Life* (Cambridge: Cambridge University Press).

Douty, Christopher. 1972. "Disasters and Charity: Some Aspects of Cooperative Economic Behavior." *American Economic Review*, Vol. 62, No. 4 (September): pp. 580–90.

Dow, Gregory. 1987. "The Function of Authority in Transaction Costs Economics." *Journal of Economic Behavior and Organization*, Vol. 8, No. 1 (March) pp. 13–38.

Doyle, Roger. 1999. "Behind Bars in the U.S. and Europe." *Scientific American*, Vol. 281, No. 2 (August): p. 25.

Draper, Hal. 1987. *The "Dictatorship of the Proletariat" from Marx to Lenin* (New York: Monthly Review Press).

Edwards, Rick. 1975. "Stages in Corporate Stability and the Risks of Corporate Failure." *Journal of Economic History,* Vol. 35, No. 2 (June): pp. 428–60.

Ehrlich, Paul R. and Anne H. Ehrlich. 1996. *Betrayal of Science and Reason: How Anti-Environmental Rhetoric Threatens Our Future* (Washington, D.C.: Island Press).

Eichengreen, Barry. 1991. "Historical Research on International Lending and Debt." *Journal of Economic Perspectives,* Vol. 5, No. 2 (Spring): pp. 149–69.

Eisner, Robert. 1985. "The Total Income System of Accounts." Survey of Current Business Vol. 65, No. 1 (January): pp. 24–48.

___. 1988. "Extended Accounts for National Income and Product." *Journal of Economic Literature,* 26: 4 (December): pp. 1611–85.

___. 1994. *The Misunderstood Economy: What Counts and How to Count It* (Boston: Harvard Business School Press).

Elkins, Stanley. 1968. *Slavery: A Problem in American Institutional and Intellectual Life* (Chicago: University of Chicago Press).

Ellickson, Robert C. 1991. *How Neighbors Settle Disputes* (Cambridge: Harvard University Press).

Emerson, Harrison. 1912. *The Twelve Principles of Scientific Management* (New York: Engineering Company).

Engels, Frederick. 1894. *Anti-Duhring: Herr Eugen Duhring's Revolution in Science* (Moscow: Progress Publishers, 1969).

English, William B. 1996. "Understanding the Costs of Sovereign Default: American State Debts in the 1840s." *American Economic Review,* Vol. 86, No. 1 (March): pp. 259–75.

Environmental Protection Agency. Office of Solid Waste. 1999. Basic Facts. <http://www.epa.gov/epaoswer/non-hw/muncpl/facts.htm>

Etzioni, Amitai. 1988. *The Moral Dimension: Toward A New Economics'* (New York: The Free Press).

Falls, Cyril. 1941. *The Nature of Modern Warfare* (New York: Oxford University Press).

Fernandez, Raquel and Richard Rogerson. 1996. "Income Distribution, Communities, and the Quality of Public Education." *Quarterly Journal of Economics,* Vol. 111, No. 1 (February): pp. 135–64.

Fisher, Franklin M., Zvi Griliches, and Carl Kaysen. 1962. "The Costs of Automobile Changes Since 1949." *Journal of Political Economy,* Vol. 70, No. 5 (October): pp. 433–51.

Fisher, Irving. 1930. *The Stock Market Crash—And After* (New York: Macmillan).

Ford, Henry, with the assistance of S. Crowther. 1922. *My Life and Work* (Garden City, NY: Garden City Publishing).

Fourier, Charles. 1835. *La Fausse Industrie.* vols. 8 and 9. *Ouvres Complètes de Charles Fourier* (Paris: Editions Anthropos).

___. 1838. *Theorie de l'Unite Universelle.* vols. 2–5 *Ouvres Complètes de Charles Fourier* (Paris: Editions Anthropos).

___. 1901. *Selections from the Works of Fourier* (London: Swan Sonnenschein).

___. 1971. *The Utopian Vision of Charles Fourier: Selected Texts on Work, Love, and Passionate Attraction,* Jonathan Beecher and Richard Bienvenu. trs. and eds. (Boston: Beacon Press).

Franklin, Benjamin. 1905–1907. *The Writings of Benjamin Franklin,* 10 vols. Albert Henry Smyth, ed. (New York: Macmillan).

Freeman, Richard B. 1996. "Why Do So Many Young American Men Commit Crimes and What Might We Do About It?" *Journal of Economic Perspectives,* Vol. 10, No. 1 (Winter): pp. 25–42.

Frey, Bruno S. 1997. *Not Just for the Money: An Economic Theory of Personal Motivation* (Brookfield, VT.: Edward Elgar).

Friedman, Milton. 1997. "Economics of Crime." *The Journal of Economic Perspectives,* Vol. 11, No. 2 (Spring): p. 194.

Frost, Robert. 1977. "Mending Wall." *North of Boston: Poems* (New York: Dobb, Mead): pp. 5–6.

Galarza, Ernesto. 1977. *Farm Workers and Agri-business in California, 1947–1960* (Notre Dame: University of Notre Dame Press).

Galbraith, John Kenneth. 1981. *A Life in Our Times* (Boston: Houghton Mifflin).

___. 1994. *A Journey Through Economic Time: A Firsthand View* (Boston: Houghton Mifflin).

Genovese, Eugene D. 1976. *Roll, Jordan, Roll: The World the Slaves Made* (New York: Vintage).

Gilligan, James. 1998. "Reflections From a Life Behind Bars: Build Colleges, Not Prisons." *Chronicle of Higher Education* (October 16): pp. B7 and B9.

Glaeser, Edward L. 1998. "Are Cities Dying?" *Journal of Economic Perspectives,* Vol. 12, No. 2 (Spring): pp. 139–60.

Goldin, Claudia and Lawrence F. Katz. 1999. "The Shaping of Higher Education: The Formative Years in the United States, 1890 to 1940." *Journal of Economic Perspectives,* Vol 13, No. 1 (Winter): pp. 37–62.

Gompers, Samuel. 1883. "Testimony," in John A. Garraty, ed. *Labor and Capital in the Gilded Age: Testimony Taken by the Senate Committee Upon the Relations Between Labor and Capital (Boston: Little, Brown, 1968).*

___. 1925. *Seventy Years of Life and Labor: An Autobiography,* 2 vols. (New York: Kelley, 1967).

Gordon, David M. 1996. *Fat and Mean: The Corporate Squeeze of Working Americans and the Myth of Managerial Downsizing* (New York: New Press).

Gould, Eric, Bruce Weinberg, and David Mustard. 1998. "Crime Rates and Local Labor Market Opportunities in the United States: 1979–1995." University of Georgia, Working Paper No. 98–472 (August).

Gray, Wayne B. and John T. Scholz. 1991. "Do OSHA Inspections Reduce Injuries: A Panel Analysis." National Bureau of Economic Research Working Paper No. 3774 (July).

Greider, William. 1997. *One World, Ready or Not: The Manic Logic of Global Capitalism* (New York: Simon & Schuster).

Grogger, Jeff. 1998. "Market Wages and Youth Crime." *Journal of Labor Economics*, Vol. 16, no. 4 (October): pp. 756–91.

Gurley, John. W. 1970. "Maoist Economic Development: The New Man in the New China." *The Center Magazine*, Vol. 3, No. 3 (May): pp. 25–33; Reprinted in Charles K. Wilber, ed. *The Political Economy of Development and Underdevelopment*, edited (New York: Random House, 1973): pp. 307–19.

Haber, Samuel. 1964. *Efficiency and Uplift* (Chicago: The University of Chicago Press).

Hagen, Piet J. 1982. *Blood: Gift or Merchandise* (New York: Alan R. Liss).

Hall, Robert E. and Charles I. Jones. 1999. "Why Do Some Countries Produce So Much More Output per Worker than Others?" *Quarterly Journal of Economics*, Vol. 114, No. 1 (February): pp. 83–116.

Hall, Robert E. and Alvin Rabushka. 1983. *Low Tax, Simple Tax, Flat Tax* (New York: McGraw-Hill).

Harberger, Arnold C. 1954. "Monopoly and Resource Allocation." *American Economic Review: Proceedings*, Vol. 44 (May): pp. 77–87.

Harrod, Roy. 1958. "The Possibility of Economic Satiety—Use of Economic Growth for Improving the Quality of Education and Leisure." in *Problems of United States Economic Development* (New York: Committee for Economic Development), Vol. 1: pp. 207–13.

Hart, Oliver. 1988. "Capital Structure as a Control Mechanism in Corporations." *Canadian Journal of Economics*, Vol. 21, No. 3 (August): pp. 467–76.

Hasek, Jaroslav. 1974. *The Good Soldier Sveijk*, Cecil Parrot, trans. (New York: Thomas Y. Crowell).

Hawley, Ellis. 1981. "Three Facets of Hooverian Associationism: Lumber, Aviation, and Movies, 1921–1930." in Thomas K. McCraw, ed. *Regulation in Perspective: Historical Essays* (Cambridge: Harvard University Press): pp. 95–123.

Hawtrey, Ralph G. 1925. *The Economic Problem* (London: Longmans, Green).

Hays, Constance L. 1999. "Variable-Price Coke Machine Being Tested." *New York Times* (28 October): p. C 1.

Hebb, Donald O. 1930. "Elementary School Methods." *Teachers' Magazine* (Montreal), Vol 12, pp. 23–6.

____. 1955. "Drives and the C.N.S." *Psychological Review*, Vol. 62, p. 246.

Henwood, Doug. 1997. *Wall Street: How It Works and for Whom* (London: Verso).

Herbert, Victor. 1999. "Short Supplies Would End If Iron-Rich Blood Weren't Trashed." *Sacramento Bee* (January 22): p. B7.

Herling, John. 1962. *The Great Price Conspiracy* (Washington, D.C.: R. B. Luce).

Higgs, Robert. 1999. "Lock 'em Up!," *Independent Review*, Vol. 4, No. 2 (Fall): pp. 309–33.

Hines, James R. Jr. 1999. "Three Sides of Harberger Triangles." *Journal of Economic Perspectives*, Vol. 13, No. 2 (Spring): pp. 167–188.

Hines, James R., Jr. and Eric M. Rice. 1994. "Fiscal Paradise: Foreign Tax Havens and American Business." *Quarterly Journal of Economics*, Vol. 109, No. 1 (February): pp. 149–82.

Hirsch, Fred. 1976. *Social Limits to Growth* (London: Routledge and Kegan Paul).

Hirschman, Albert O. 1977. *The Passions and the Interests: Political Arguments for Capitalism Before Its Triumph* (Princeton: Princeton University Press).

Hirschman, Albert. 1984. "Against Parsimony: Three Easy Ways of Complicating Some Categories of Economic Discourse." *American Economic Review*, Vol. 74, No. 2 (May): pp. 89–95.

___. 1987. *Economic Behavior in Adversity* (Chicago: University of Chicago Press).

Hobbes, Thomas. 1651. *Leviathan*, C. B. Macpherson, ed. (Baltimore: Penguin Books, 1968).

Hobsbawm, E. 1986. *Labouring Men: Studies in the History of Labour* (London: Weidenfeld and Nicolson).

Hoover, Herbert. 1921. "Foreword." American Engineering Council. *Waste in Industry by the Committee on Elimination of Waste in Industry of the Federated American Engineering Societies*. 1st ed. (Washington, D.C.: Federated American Engineering Societies; NY: McGraw-Hill).

Horn, Joshua S. 1971. *Away with All Pests: An English Surgeon in People's China, 1954–1969* (New York: Monthly Review Press).

Huizinga, Johan. 1970. *Homo Ludens: A Study of the Play Element in Culture* (New York: Harper and Row).

Hume, David. 1742. "Of Independency of Parliament." *Essays. Moral, Political, and Literary*, vol. 1 (Oxford: Oxford University Press, 1963): pp. 117–18.

Huston, John and Nipoli Kamdar. 1996. "$9.99: Can "Just-Below" Pricing Be Reconciled with Rationality?" *Eastern Economic Journal*, Vol. 22, No. 2 (Spring): pp. 137–45.

Hutchins, Robert. 1944. "Threat to American Education." *Collier's*, No. 114 (December 30): pp. 20–21.

Jacobs, Jane. 1961. *The Death and Life of Great American Cities* (New York: Random House).

Jacoby, Russell. 1994. *Dogmatic Wisdom: How the Culture Wars Divert Education and Distract America* (New York: Anchor Books).

Jenkins, Holman W., Jr. 1998. "The Rise and Stumble of Nike." *Wall Street Journal* (June 3): p. A19.

Jensen, Michael C. 1986. "Agency Costs of Free Cash Flow, Corporate Finance, and Takeovers." *American Economic Review*, Vol. 76, No. 2 (May): pp. 323–9.

___. 1988. "Takeovers: Their Causes and Consequences." *Journal of Economic Perspectives*, Vol. 2, No. 1 (Winter): pp. 21–48.

___. 1993. "The Modern Industrial Revolution: Exit and the Failure of Internal Control Systems." *Journal of Finance*, Vol. 48, No. 3 (July): pp. 831–80.

Jensen, Michael and William Meckling. 1976. "Theory of the Firm: Managerial Behavior, Agency Costs, and Ownership Structure." *Journal of Financial Economics,* Vol. 3, pp. 305–60; reprinted in part in Louis Putterman, ed. *The Economic Nature of the Firm: A Reader* (Cambridge: Cambridge University Press, 1986): pp. 209–29.

Jewkes, John, David Sawyers, and Richard Stillerman. 1958. *The Sources of Invention* (London: Macmillan).

Johnson, Harry G. 1975. *On Economics and Society* (Chicago: University of Chicago Press).

Karpoff, Jonathan M., D. Scott Lee, and Valaria P. Vendrzyk. 1999. "Defense Procurement Fraud, Penalties, and Contractor Influence." *Journal of Political Economy,* forthcoming. Probably Vol. 107, No. 5 (October).

Kaufman, Jonathan. 1998. "They Know They're Well Off, But They Can't Help Coveting." *Wall Street Journal* (August 3).

Kay, Jane. 1997. *Asphalt Nation: How the Automobile Took Over America, And How We Can Take It Back* (New York: Crown).

Keller, Edmund R. 1977. "The Trouble with Oligopoly Is the Price." *Antitrust Law and Economics Review,* Vol. 9, No. 2, pp. 73–91.

Keynes, John Maynard, 1930. "Economic Possibilities for Our Grandchildren." *Nation and Athenaeum* (11 and 18 October); in *Essays in Persuasion,* Vol. 9 of *The Collected Works of John Maynard Keynes* (London: Macmillan): pp. 321–31.

Kidder, Tracy. 1981. *The Soul of a New Machine* (Boston: Little, Brown).

Kiester, Edwin, jr. 1994. "The GI Bill May Be the Best Deal Ever Made by Uncle Sam." *Smithsonian Magazine,* Vol. 25, No. 4 (November): pp. 129–32.

King, John E. 1983. "Utopian or Scientific? A Reconsideration of the Ricardian Socialists." *History of Political Economy,* Vol. 15, No. 3 (Fall): pp. 345–73.

Knoedler, Janet T. 1997. "Veblen and Technical Efficiency." *Journal of Economic Issues,* Vol. 31, No. 4 (December): pp. 1011–26.

Knoedler, Janet and Anne Mayhew. 1994. "The Engineers and Standardization." *Business and Economic History,* Vol. 23, No. 1 (Fall): pp. 141–51.

Kolko, Gabriel. 1990. *The Politics of War: The World and United States Foreign Policy, 1943–1945* (New York: Pantheon Books).

Krueger, Anne O. 1974. "The Political Economy of the Rent-Seeking Society." *American Economic Review,* Vol. 64, No. 4 (June): pp. 291–303.

Kuttner, Robert. 1997. *Everything For Sale: The Virtues and Limits of Markets* (New York: Alfred A. Knopf).

Laband, David N. and John P. Sophocleus. 1992. "An Estimate of Resource Expenditures on Transfer Activity in the United States." *Quarterly Journal of Economics,* Vol. 107, No. 3 (Autumn): pp. 959–83.

Landes, David S. 1998. *The Wealth and Poverty of Nations: Why Some Are So Rich and Some So Poor* (New York: W. W. Norton).

Layton, Edwin T., Jr. 1971. *The Revolt of the Engineers* (Cleveland: The Press of Case Western Reserve University).

Lefebvre, Henri. 1971. *Everyday Life in the Modern World,* Sacha Rabinovitch, trans.(New York: Harper & Row).

Leibenstein, Harvey. 1966. "Allocative Efficiency and X-Efficiency." *American Economic Review,* Vol. 56: pp. 392–415; in Louis Putterman, ed. *The Economic Nature of the Firm: A Reader* (Cambridge: Cambridge University Press, 1986): pp. 165–9.

Leigh, J. Paul, Steven B. Markowitz, Marianne Fahs, Chonggah Shin, and Philip J. Landrigan. 1997. "Occupational Injury and Illness in the United States: Estimates of Costs, Morbidity and Mortality." *Archives of Internal Medicine,* No. 167 (July): pp. 1557–68.

Lemieux, Thomas and David Card. 1998. "Education, Earnings, and the "Canadian G.I. Bill." (June) unpub.

Lewis, Michael. 1989. *Liar's Poker: Rising Through the Wreckage on Wall Street* (New York: W.W. Norton).

Lippert, John. 1978. "Shopfloor Politics at Fleetwood." *Radical America,* Vol. 12, No. 4 (July/August) pp. 25–44.

Locke, John. 1663. *Essays on the Law of Nature,* W. von Leyden, ed. (Oxford: Clarendon Press, 1954).

Lucas, Robert E., Jr. 1970. "Capacity, Overtime, and Empirical Production Functions." *American Economic Review,* Vol. 60, No. 2 (May): pp. 23–27; reprinted in his *Studies in Business Cycle Theory* (Cambridge, MA: MIT Press, 1983): pp. 146–55.

Macaulay, Thomas Babington. 1840. "Lord Clive." *Edinburgh Review* (January); reprinted in G. M. Young, ed. *Macaulay: Prose and Poetry,* 7 vols. (Cambridge: Harvard University Press): vol. 6, pp. 306–72.

McGrane, Reginald C. 1935. *Foreign Bondholders and American State Debts* (New York: MacMillan).

Madrick, Jeff. 1998. "Computers: Waiting for the Revolution." *New York Review of Books,* Vol. 45, No. 4 (March 26): pp. 29–33.

Magee, Stephen, William Brock, and Leslie Young. 1989. *Black Hole Tariffs and the Endogenous Policy Theory* (Cambridge: Cambridge University Press).

Maital, Shlomo. 1982. *Minds, Markets, and Money: Psychological Foundations of Economic Behavior* (New York: Basic Books).

Mankiw, N. Gregory and Michael D. Whinston. 1986. "Free Entry and Social Efficiency." *Rand Journal of Economics,* Vol. 17, No. 1 (Spring): pp. 48–58.

Marshall, Alfred. 1889. "Cooperation." in Alfred C. Pigou, ed. *Memorials of Alfred Marshall* (New York: Kelley and Millman, 1956; 1st ed, 1925): pp. 227–55.

___. 1923. *Industry and Trade: A Study of Industrial Technique and Business Organization; And of their Influences on the Conditions of Various Classes and Nations,* 4th ed. (London: Macmillan & Co.; New York: Augustus M. Kelley, 1970).

___. 1927. *Principles of Economics: An Introductory Volume,* 8th ed. (London: MacMillan and Co.).

Marx, Karl. 1847. *The Poverty of Philosophy* (New York: International Publishers, 1963).

____. 1979. *Revelations Concerning the Communist Trial in Cologne.* in Marx and Engels, *Collected Works,* Vol. 11. Marx and Engels: 1851–53. (New York: International Publishers): pp. 395–497.

____. 1977. *Capital,* Vol. 1 (New York: Vintage).

____. 1963–1971. *Theories of Surplus Value,* 3 Parts (Moscow: Progress Publishers).

____. 1981. *Capital,* vols. 2 and 3 (New York: Vintage).

Marx, Karl and Frederick Engels. 1846. *The German Ideology,* in *Karl Marx and Friedrich Engels, Collected Works.* Vol. 5. *Marx and Engels: 1845–1847* (New York: International Publishers, 1976): pp. 19–540.

Mathewson, Stanley. 1939. *Restriction of Output Among Unorganized Workers* (Carbondale: Southern Illinois University Press, 1969).

McCloskey, Donald and Arjo Klamer. 1995. "One Quarter of the GDP is Persuasion." *American Economic Review,* Vol. 85, No. 2 (May): pp. 191–5.

Meyer, Stephen. 1981. *The Five Dollar Day: Labor Management and Social Control in the Ford Motor Company, 1908–1921* (Albany: State University of New York Press).

Mill, John Stuart. 1848. *Principles of Political Economy with Some of Their Applications to Social Philosophy,* vols. 2–3. *Collected Works,* J. M. Robson, ed. (Toronto: University of Toronto Press).

Miller, James P. et al. 1995. "Bad Chemistry: W.R. Grace is Roiled By Flap Over Spending and What to Disclose." *Wall Street Journal* (March 10): pp. A 1 and A 16.

Miller, Jerome G. 1996. *Search and Destroy: African-American Males in the Criminal Justice System* (Cambridge: Cambridge University Press).

Minsky, Terry. 1981. "Gripes of Rath: Workers Who Bought Iowa Slaughterhouse Regret That They Did." *Wall Street Journal* (2 December).

Mohammad, Sharif, and John Whalley. 1984. "Rent Seeking in India: Its Costs and Policy Significance." *Kyklos,* Fasc. 37, No. 3: pp. 387–413.

Mokhiber, Russell. 1988. *Corporate Crime and Violence: Big Business Power and the Abuse of the Public Trust* (San Francisco: Sierra Club Books).

Monroe, Alan. 1998. "Public Opinion and Public Policy, 1980–1993." *Public Opinion Quarterly,* 62: 1 (Spring): pp. 6–28.

Morgenstern, Oskar. 1972. "Thirteen Critical Points in Contemporary Economic Theory: An Interpretation." *Journal of Economic Literature,* Vol. 10, No. 4 (December) pp. 1163–89.

Morton, Fiona and Joel Podolny. 1998. "Love or Money? The Effects of Owner Motivation in the California Wine Industry." National Bureau of Economic Research Working Paper No. 6743 (October).

Moschandreas, Maria. 1997. "The Role of Opportunism in Transaction Cost Economics." *Journal of Economic Issues,* Vol. 31, No. 1 (March): pp. 39–58.

Moseley, Fred. 1991. *The Falling Rate of Profit in the Postwar United States Economy* (New York: St. Martin's Press).

Mukerjee, M. 1994. "Wall Street: Refugees from Physics Find Joy as 'Derivatives Geeks'." *Scientific American,* Vol. 271, No. 4 (October): pp. 126 and 128.

Mulligan, Casey B. 1998. "Pecuniary Incentives to Work in the United States during World War II." *Journal of Political Economy,* Vol. 106, No. 5 (October): pp. 1033–77.

Myrdal, Gunnar. 1962. *An American Dilemma: The Negro Problem and Modern Democracy* (New York: Harper and Row).

Naples, Michele I. 1988. "Industrial Conflict, the Quality of Worklife, and the Productivity Slowdown in U.S. Manufacturing." *Eastern Economic Journal,* Vol. 14, No. 1 (January–March) pp. 157–66.

National Industrial Conference Board. 1929. *Industrial Standardization* (New York: National Industrial Conference Board).

National Safety Council. 1998. "Accident Facts." <http://www.nsc.org/lrs/statinfo/af98spk.htm>.

Nelson, Cary and Michael Berube. 1994. "Graduate Education Is Losing Its Moral Base." *The Chronicle of Higher Education* (March 23) p. B1.

Neulinger, John. 1981. *The Psychology of Leisure,* 2d ed. (Springfield, IL: Charles C. Thomas).

New York City. Department of Sanitation. 1999. *Fact Sheet.* <http://www.ci.nyc.ny.us/html/dos/html/dosfact.html>.

Niggle, Christopher J. 1988. "The Increasing Importance of Financial Capital in the U.S. Economy." *Journal of Economic Issues,* Vol. 22, No. 2 (June): pp. 581–8.

Noble, David. 1979. *America By Design* (New York: Oxford University Press).

___. 1984. *Forces of Production: A Social History of Automation* (New York: Oxford University Press).

Norsworthy, J. R. and Craig Zabala. 1985. "Worker Attitudes, Worker Behavior, and Productivity in the U.S. Automobile Industry, 1959–1976." *Industrial and Labor Relations Review,* Vol. 38, No. 4 (July): pp. 544–57.

North, Douglass Cecil. 1990. *Institutions, Institutional Change, and Economic Performance* (Cambridge: Cambridge University Press).

Olson, Keith. 1974. *The G.I. Bill, the Veterans, and the Colleges* (Lexington, KY: University Press of Kentucky).

Orr, Douglas V. 1998. "Strategic Bankruptcy and Private Pension Default." *Journal of Economic Issues,* Vol. 32, No. 3 (September): pp. 669–87.

O'Toole, James. 1981. *Making America Work: Productivity and Responsibility* (New York: Continuum).

Palmer, Bryan. 1975. "Class, Conception and Conflict: The Thrust for Efficiency, Managerial Views of Labor and the Working Class Rebellion, 1903–22." *Review of Radical Political Economy,* Vol. 7, No. 2 (Summer): pp. 31–50.

Parenti, Christian. 1999. *Lockdown America: Police and Prisons in the Age of Crisis* (New York: Verso).

Parker, Suzi. 1998. "Blood Money." *Salon Magazine* (December 23) http://www.salon1999.com/news/1998/12/cov_23news.html.

Pearson, Hesketh. 1934. *The Smith of Smiths* (London: Hamish Hamilton).

Perelman, Michael A. 1977. *Farming for Profit in a Hungry World: Capital and the Crisis in Agriculture* (Totowa, NJ: Allenheld, Osmun).

___. 1989. "Adam Smith and Dependent Social Relations." *History of Political Economy,* Vol. 21, No. 3 (Fall): pp. 503–20.

___. 1993. *The Pathology of the U.S. Economy: The Costs of a Low Wage System* (NY and London: St. Martin's and Macmillan).

___. 1996. *The End of Economics* (London: Routledge).

___. 1998. *Class Warfare in the Information Age* (New York: St. Martin's Press).

___. 1999. *The Natural Instability of Markets: Expectations, Increasing Returns and the Collapse of Markets* (New York: St. Martin's Press).

___. 2000. *The Invention of Capitalism: The Secret History of Primitive Accumulation* (Durham: Duke University Press).

Persson, Torsten and Guido Tabellini. 1994. "Is Inequality Harmful for Growth?" *American Economic Review,* Vol. 84, No. 3 (June): pp. 600–21.

Phillips, Kevin. 1994. *Arrogant Capital: Washington, Wall Street, and the Frustration of American Politics* (Boston: Little, Brown).

Pinker, Steven. 1997. *How the Mind Works* (New York: Norton).

Ponson, T. 1854. *Traite de l'exploitation des mines de houille* (Liege).

Posner, Richard A. 1975. "The Social Costs of Monopoly and Regulation." *Journal of Political Economy,* Vol. 83, No. 4 (August): pp. 807–27.

Prince, S. H. 1920. *Catastrophe and Social Change* (New York: Columbia University Press).

Public Citizen. 1998. The Facts About Products Liability Lawsuits. <http://www.citizen.org/congress/civjus/product/facts1.htm>

Putnam, Robert, 1995. "Bowling Alone." *Journal of Democracy,* Vol. 6, No. 1 (January): pp. 65–78.

Raff, Daniel M. G. and Summers, Lawrence H. 1987. "Did Henry Ford Pay Efficiency Wages?' *Journal of Labor Economics,* Vol. 5, No. 4 (Part 2) (October): pp. S57–S86.

Ravenstone, Piercy. 1821. *A Few Doubts as to the Correctness of Some Opinions Generally Entertained on the Subjects of Population and Political Economy* (New York: Kelley, 1966): p. 221.

Reiman, Jeffrey H. 1996. *And the Poor Get Prison: Economic Bias In American Criminal Justice* (Boston: Allyn and Bacon).

Ricardo, David. 1821. *The Principles of Political Economy and Taxation,* vol. 1. *The Works and Correspondence of David Ricardo,* 11 vols. Piero Sraffa, ed. (Cambridge: Cambridge University Press, 1951–1973).

Roberts, Susan M. 1995. "Small Place, Big Money: The Cayman Islands and the International Financial System." *Economic Geography,* Vol. 7, No. 3 (July): pp. 237–56.

Rosenberg, Nathan. 1990. "*Adam Smith and the Stock of Moral Capital.*" *History of Political Economy,* Vol. 22, No. 1 (Spring): pp. 1–18.

Rosenthal, A. M. and Arthur Gelb. 1965. *The Night the Lights Went Out* (New York: Signet).

Rustad, Michael. 1998. "Unraveling Punitive Damages: Current Data and Further Inquiry." *Wisconsin Law Review,* No. 1 (Spring): pp. 15–69.

Sabel, Charles F. 1993. "Studied Trust: Building New Forms of Cooperation in a Volatile Economy." *Human Relations,* Vol. 46, No. 9, pp. 1133–70.

Sampson, Anthony. 1982. *The Money Lenders: Bankers and a World in Turmoil* (New York: Viking Press).

Samuels, Warren and Edward Puro. 1991. "The Problem of Price Controls at the Time of Natural Disaster." *Review of Social Economy,* Vol. 49, No. 1 (Spring): pp. 62–75.

Sapolsky, Robert M. 1994. *Why Zebras Don't Get Ulcers: A Guide to Stress, Stress-Related Diseases, and Coping* (New York: W. H. Freeman and Company).

Scherer, F. M. 1976. "Industrial Structure, Scale Economies, and Worker Alienation." in Robert T. Masson and P. David Qualls, eds. *Essays on Industrial Organization in Honor of Joe S. Bain* (Cambridge, MA: Ballinger Publishing Co.): pp. 105–21.

Schor, Juliet B. 1991. *The Overworked American: The Unexpected Decline of Leisure* (New York: Basic Books).

Schweickart, David. 1996. *Against Capitalism* (Boulder, CO: Westview Press).

Scitovsky, Tibor. 1976. *The Joyless Economy: An Inquiry into Human Satisfaction* (New York: Oxford University Press).

___. 1991. "Hindsight Economics." *Banca Nazionale del Lavoro Quarterly Review,* No. 178 (September): pp. 251–70.

Scott, James C. 1998. *Seeing Like a State: How Certain Schemes to Improve the Human Condition Have Failed* (New Haven: Yale University Press).

Sen, Amartya. 1973. "Behaviour and the Concept of Preference." *Economica,* Vol. 40, No. 159 (August): pp. 241–59.

Senior, Nassau W. 1836. *An Outline of the Science of Political Economy,* 3d ed. (New York: A. M. Kelly, 1951).

Shaiken, Harley. 1985. *Work Transformed: Automation and Labor in the Computer Age* (New York: Holt, Rinehart & Winston).

Shaikh, Anwar and E. Ahmet Tonak. 1994. *Wealth of Nations: The Political Economy of National Accounts* (Cambridge: Cambridge University Press).

Silverstein, Ken. 1998. *Washington on $10 Million a Day: How Lobbyists Plunder the Nation* (Monroe, ME: Common Courage Press).

Simon, Herbert Alexander. 1979. "Rational Decision Making in Business Organizations." *American Economic Review,* Vol. 69, No 4 (September): pp. 493–513.

___. 1985. "Human Nature in Politics: The Dialogue of Psychology with Political Science." *American Political Science Review,* Vol. 79, No. 2 (June): pp. 293–304.

Skidelsky, Robert. 1992. *John Maynard Keynes.* Vol. 2. *The Economist as Savior: 1920–1937* (London: Macmillan).

Skocpol, Theda. 1998. "The G.I. Bill and U.S. Social Policy, Past and Future," in Ellen Frankel Paul, Fred D. Miller Jr., Jeffrey Paul, eds. *The Welfare State* (Cambridge: Cambridge University Press): pp. 95–115.

Smith, Adam. 1759. *The Theory of Moral Sentiments,* D. D. Raphael and A. L. Macfie, eds. (Oxford: Clarendon Press, 1976).

____. 1776. *An Inquiry into the Nature and Causes of the Wealth of Nations,* 2 vols. R. H. Campbell and A. S. Skinner, eds. (New York: Oxford University Press, 1976).

____. 1978. *Lectures on Jurisprudence,* R. L. Meek, D. D. Raphael, and P. G. Stein, eds. (Oxford: Clarendon University Press).

Smith, Richard Alan. 1961. "The Incredible Electrical Conspiracy." *Fortune* (April).

Smith, Sidney. 1809. "Review of R. L. Edgeworth. *Essays on Professional Education.*" *Edinburgh Review,* Vol. 29 (October): pp. 40–51; quoted as "Utility and Anti-Clacissism, 1809," in Michael Sanderson, ed. *The Universities in the Nineteenth Century* (London: Routledge and Kegan Paul, 1975): p. 35.

Solow, Robert S. 1971. "Blood and Thunder." *Yale Law Journal,* Vol. 80, No. 8 (July): pp. 1696–1711.

Starr, Douglas. 1998. *Blood: An Epic History of Medicine and Commerce* (New York: Knopf).

Stauber, John C. and Sheldon Rampton. 1995. *Toxic Sludge Is Good for You!: Lies, Damn Lies and the Public Relations Industry* (Monroe, ME: Common Courage Press).

Stephan, Paula. 1996. "The Economics of Science." *Journal of Economic Literature,* Vol. 34, No. 3 (September): pp. 1199–1262.

Stephan, Paula E. and Sharon G. Levin. 1992. *Striking the Mother Lode in Science: The Importance of Age, Place, and Time* (New York: Oxford University Press),

Stonebraker, Robert J. 1979. "Turnover and Mobility among the 100 Largest Firms: An Update." *American Economic Review,* Vol. 69, No. 5 (December): pp. 968–73.

Strauss, E. 1954. *Sir William Petty, Portrait of a Genius* (Glencoe, IL: The Free Press).

Summers, Lawrence H. and Victoria P. Summers. 1989. "When Financial Markets Work too Well: A Cautious Case for a Securities Transactions Tax." *Journal of Financial Services Research,* Vol. 3, Nos. 2 and 3 (December): pp. 261–86.

Sunstein, Cass R. 1997. *Free Markets and Social Justice* (Oxford: Oxford University Press).

Sward, Keith. 1972. *The Legend of Henry Ford* (New York: Atheneum).

Tawney, Richard H. 1929. *Equality* (New York: Harcourt, Brace and World).

Thompson, George. 1954. "Intercompany Technical Standardization in the Early American Automobile Industry." *Journal of Economic History,* Vol. 14, No. 1 (Winter): pp. 1–20.

Thurow, Lester C. 1971. "The Income Distribution as a Pure Public Good." *Quarterly Journal of Economics,* Vol. 85, No. 2 (May): pp. 327–336.

____. 1998. "Wage Dispersion: 'Who Done It'?" *Journal of Post Keynesian Economics,* Vol. 21, No. 1 (Fall): pp. 25–37.

Tillinghast-Towers Perrin. 1995. *Tort Cost Trends: An International Perspective* (NY).

Titmuss, Richard M. 1958. "War and Social Policy." *Essays on the Welfare State,* Richard Titmuss, ed. (London: Allen and Unwin): pp. 75–87.

____. 1971. *The Gift Relationship: From Human Blood to Social Policy* (New York: Random House).

Townsend, Joseph. 1786. *A Dissertation on the Poor Laws by a Well Wisher to Mankind*. Reprinted in John R. McCulloch, ed. *A Select Collection of Scarce and Valuable Economic Tracts* (New York: Augustus M. Kelley, 1966): 395–450.

Triebel, Armin. 1997. "Coal and the Metropolis," in Jay Winter and Jean-Louis Robert, eds. *Capital Cities at War: London, Paris, Berlin, 1914–1919* (Cambridge and New York: Cambridge University Press): pp. 342–73.

Tsuru, Shigeto. 1993. *Japan's Capitalism: Creative Defeat and Beyond* (Cambridge: Cambridge).

United States Commissioner of Labor. 1904. *Eleventh Special Report of the Commissioner of Labor, Regulation and Restriction of Output* (Washington D.C., 1904).

United States Congress. Joint Economic Committee. 1996. *Improving the American Legal System: The Economic Benefits of Tort Reform* (Washington, D.C.: U.S. Government Printing Office).

United States Constitutional Convention. 1787. *Notes of Debate in the Federal Convention of 1787, reported by James Madison* (Athens: Ohio University Press, 1966).

United States Department of Commerce. National Telecommunications and Information Administration. 1998. *Falling Through the Net. II. New Data on the Digital Divide*. <http://www.ntia.doc.gov/ntiahome/net2/>.

United States Department of Justice, Bureau of Justice Statistics. 1997. *Sourcebook of Criminal Justice Statistics* (Washington, D.C.: United States Department of Justice).

United States Department of Justice. 1998. *Profile of Jail Inmates 1996* (Washington, D.C.: U.S. Department of Justice, Office of Justice Programs, NCJ-164620, April).

United States Environmental Protection Agency. 1999. *Municipal Solid Waste Factbook*, <www.epa.gov/epaoswer/non-hw/muncpl/factbook/internet/mswf/gen.htm#27>.

United States House of Representatives, Committee on Education and Labor, Subcommittee on Elementary, Secondary and Vocational Education. 1989. *Child Nutrition Programs: Issues for the 101st Congress,* One Hundredth Congress, 2d Session (December).

Ure, Andrew. 1835. *The Philosophy of Manufactures* (London); cited in Marx 1977.

Vanek, Jaroslav. 1989. *Crisis and Reform: East and West* (Ithaca, NY: Author).

Veblen, Thorstein. 1899. *The Theory of the Leisure Class: An Economic Study of Institutions* (New York: Mentor Books, 1953).

Veblen, Thorstein. 1904. *The Theory of Business Enterprise* (Clifton, NJ: Augustus M. Kelley, 1975).

____. 1921. *The Engineers and the Price System* (New York: Viking Press, 1938; NY: A. M. Kelley, 1965).

____. 1934. "Farm Labor for the Period of the War." *Essays in Our Changing Order,* Leon Ardzrooni, ed. (New York: Viking Press, 1954): pp. 279–318.

Wallis, J. J. and Douglass C. North. 1986. "Measuring Transaction Costs in the American Economy, 1870–1970." in Stanley L. Engerman and Robert G. Gallman, eds. *Long Term Factors in American Economic Growth*. National Bureau of

Economic Research. Studies in Income and Wealth, vol. 51 (Chicago: University of Chicago Press): pp. 94–148.

Watson, Bill. 1971. "Counter-Planning on the Shop Floor." *Radical America*, Vol. 5, No. 3 (May-June): pp. 77–85.

Welch, Finis. 1999. "In Defense of Inequality." *American Economic Review*, Vol. 89, No. 2 (May): pp. 1–17.

Wicksteed, Philip H. 1910. "The Common Sense of Political Economy." Reprinted in *The Common Sense of Political Economy and Selected Papers and Reviews on Economic Theory*, 2 vols., Lionel Robbins, ed. (London: Routledge and Kegan Paul, 1933): ii.

Wilkins, Mira. 1989. *The History of Foreign Investment in the United States to 1914* (Cambridge: Harvard University Press).

Wilkinson, Richard G. 1997. *Unhealthy Societies: The Afflictions of Inequality* (London: Routledge).

Williamson, Oliver. 1980. "The Organization of Work: A Comparative Institutional Assessment." *Journal of Economic Behavior and Organization*, Vol. 1, No. 1 (March): pp. 5–38; partially reprinted in Louis Putterman, ed. *The Economic Nature of the Firm: A Reader* (Cambridge: Cambridge University Press, 1986): pp. 292–311.

___. 1985. *The Economic Institutions of Capitalism* (New York: Macmillan).

___. 1984. "The Economics of Governance: Framework and Implications." *Journal of Institutional and Theoretical Economics*, Vol. 140, No. 1 (March): pp. 195–233.

___. 1985. *The Economic Institutions of Capitalism: Firms, Markets, Relational Contracting* (New York: The Free Press).

___. 1987. "Transaction Cost Economics: The Comparative Contracing Perspective." *Journal of Economic Behavior and Organization*, Vol. 8, No. 4 (December): pp. 617–25.

Wills, Gary. 1982. *Explaining America: The Federalist* (New York: Penguin Books).

Wilson, Joan Hoff. 1975. *Herbert Hoover, Forgotten Progressive* (Boston: Little, Brown).

Winslow, Ron. 1998. "Casual Drinkers Are Seen As Big Productivity Problem." *Wall Street Journal* (22 December).

Winston, Gordon C. 1999. "Subsidies, Hierarchy and Peers: The Awkward Economics of Higher Education." *The Journal of Economic Perspectives*, Vol. 13, No. 1 (Winter): pp. 13–36.

Winter, Jay and Jean-Louis Robert, eds. 1997. *Capital Cities at War: London, Paris, Berlin, 1914–1919* (Cambridge and New York: Cambridge University Press).

Wood, David, Neal Halfon, Debra Scarlata, Paul Newacheck, and Sharon Nessim. 1993. "Impact of Family Relocation on Children's Growth, Development, School Function and Behavior." *Journal of the American Medical Association*, Vol. 270, No. 11 (September 15): pp. 1334–38.

Woodham-Smith, Cecil Blanche Fitz Gerald. 1951. *Florence Nightingale, 1820–1910* (New York: McGraw-Hill).

Woolhandler, Steffie, David U. Himmelstein, James P. Lewontin. 1993. "Administrative Costs in U.S. Hospitals." *The New England Journal of Medicine*, Vol. 329, No. 6 (5 August): pp. 400–3.

Young, Allyn. 1990. "Big Business: How the Economic System Grows and Evolves Like a Living Organism." *Journal of Economic Studies,* Vol. 17, Nos. 3–4: pp. 161–70; first published anonymously as chapter 38 of *The Book of Popular Science,* Vol. 15 (New York, 1929).

Zachary, G. Pascal. 1996. "Manpower to Offer Physicists as Temps." *Wall Street Journal* (November 27): pp. A2 and A14.

Index